MYSTERIES OF
LIFE AND THE UNIVERSE

NEW ESSAYS FROM

AMERICA'S FINEST

WRITERS ON SCIENCE

EDITED BY

WILLIAM H. SHORE

HARCOURT BRACE JOVANOVICH, PUBLISHERS
New York   San Diego   London

Library of Congress Cataloging-in-Publication Data
Mysteries of life and the universe: new essays from America's finest writers
on science/edited by William H. Shore.
p.   cm.
ISBN 0-15-163972-8
1. Science—Miscellanea.   I. Shore, William H.
Q173.M984   1992
500—dc20   92-15677

CONTENTS

—

CONTENTS

—

ACKNOWLEDGMENTS

The remarkable generosity of twenty-nine distinguished writers made this unique anthology possible. I am deeply grateful they were willing to donate their work to this volume so proceeds could be used by Share Our Strength to benefit children in need. They sought nothing in return except the satisfaction of helping others. Collectively their work adds up to much more than a book. It is a gift, a model for others who would give something back to their community, and an inspiration.

These pages would still be in manuscript form, scattered hopelessly among twenty-nine manila envelopes, were it not for the vision, talent, and patience of Alane Mason, my editor at Harcourt Brace Jovanovich. Her faith in this project from the very beginning and her dedication to seeing it through were invaluable. She is a good friend, and like any good friend she has been a good teacher as well. I would never have met Alane Mason if not for Flip Brophy, my agent at Sterling Lord Literistic, who has introduced me to so many and so much. She has been one of the strongest forces behind SOS's publishing efforts, and I've instructed our auditors to note, among the assets on our balance sheet, that her value is inestimable.

Marie Nash of the SOS staff was my editorial assistant on this project, or perhaps I was hers. Either way, she successfully reeled in each and every essay before deadline and deserves enormous credit for always being able to intuit what needed to be done, especially during extensive periods when other business kept me far from the office.

The Echoing Green Foundation, under the leadership of Ed Cohen and Jennifer Eplett Reilly, has helped to bring this, among many other worthy projects, to completion. The generous support of the Sloan Foundation was also instrumental, and greatly appreciated.

Finally, a word of special thanks to the rest of the SOS staff: Jennifer Hadley, Steve Kessel, Patrick Lemmon, Karen Napoli, Marisa Nightingale, Mark Risch, Harriet Robinson, Colleen Roney, Debbie Shore, and Catherine Townsend. Day in and day out, their tireless efforts have touched lives in virtually every community of America. If Thoreau was right, that the highest of the arts is "to affect the quality of the day," then together they have executed a series of true masterpieces.

W. H. S.

INTRODUCTION

Albert Einstein, the father of twentieth-century science, once said, "A person first starts to live when he can live outside himself, when he can have as much regard for his fellow man as he does for himself. Life is a gift and if we agree to accept it we must contribute in return. When we fail to contribute we fail to adequately explain why we are here." This book is irrefutable evidence that that noble spirit still lives on among those who have devoted their lives to expanding the frontiers of science.

Nearly every one of the essays gathered here has been especially written for and donated to Share Our Strength, a nonprofit organization I founded in late 1984 to raise awareness of and funds to relieve hunger, homelessness, and illiteracy, as well as to address closely related problems like infant mortality. When you think about how best to raise funds for such causes, the names of scientists and science writers like Jim Farlow, a paleontologist who has studied fossils in the Paluxy riverbed of Texas, psychology professor John Kotre, cosmologist Alan Lightman, neurologist Harold Klawans, or any of the other contributing writers may not be the first that come to mind. But each of these responded enthusiastically to our invitation to write about his or her work. And each generously donated that writing to Share Our Strength.

This unprecedented collaboration among the nation's leading scientists and science writers embodies the central philosophy of SOS: Every individual possesses unique skills that can be used in the fight against hunger. Creative individuals respond to the challenge of giving something back to their community through their skills and their art. Such a contribution of one's time and talent forges a personal commitment and a lasting connection to the cause.

This philosophy has helped Share Our Strength become the largest

national private hunger-relief foundation in the United States. What began as an effort to organize restaurants and the food-service industry to
fight hunger has now expanded to include novelists, booksellers, pho-
tographers, scientists, entertainers, and many of America's top corporate
executives. Since 1990, SOS has distributed more than seven million
dollars to food banks, homeless shelters, and training programs in nearly
a hundred cities and towns across the U.S., and to relief and development
projects in more than a dozen countries around the world.

Hunger continues to be a problem of epidemic proportions in the
U.S. It takes a severe toll upon millions of our citizens, manifesting itself
in nutritional deficiencies that wreak significant damage to their growth,
attention span, concentration, and resistance to disease. Because hunger
is symptomatic of other issues that make up the web of poverty in America
today, SOS commits a portion of its funds to addressing the broader
problems. A large portion of the proceeds from this book, for example,
will be used by SOS to support programs of the National Commission
on Infant Mortality specifically designed to improve the nutrition of preg-
nant women and young mothers, thereby giving their babies the healthiest
possible start in life. The United States has a higher rate of infant mortality
than twenty-one other countries—including Singapore, Hong Kong, and
Germany—and its percentage of babies with perilously low birth-weights
is actually increasing. I hope this anthology will raise not only funds but
also awareness of their plight.

The shape of this book evolved in a surprising and unpredictable
fashion. We did not assign any topics to our contributing writers. We
simply wanted them to write within their areas of expertise on subjects
that intrigued them. So we gave them the theme of Mysteries of Life and
the Universe and the widest possible latitude. To our surprise and great
pleasure, not only did we end up with essays covering an impressive
diversity of subjects, but those essays seemed to fit naturally together in
a symbiotic whole that is greater than the sum of each of the parts. The
variety of distinctly individual voices forms a choral symphony expressing
the excitement, wonder, and humanity of the scientific endeavor.

Whether contemplating the gentle descent of a tree's leaf toward one's feet or the faint twinkling of a distant galaxy, whether comparing theories about the disappearance of Neanderthal man or daring to guess the limits of artificial intelligence, these essays reflect on science's restless, relentless voyage through the centuries and continue its quest in attempting to solve today's most puzzling mysteries.

Unlike a rigidly structured science textbook, this collection resembles a river that meanders casually but purposefully from its source to its delta, flowing over and through diverse topics along the way, carrying the reader from the birth of the universe, through the evolution of life, then brushing up against the banks of consciousness, thought, and memory. Further downstream, one will come to some of the ways in which intelligence is applied in the pursuit of scientific objectives and, finally, to speculation about the ultimate fate of the earth and our place in the larger scheme of things.

Taken together, these essays reinforce science's great and eternal paradoxes: that knowledge leads to questions; that the more precisely we quantify, the closer we come to the immeasurable; that our discoveries today will almost certainly enable future generations to disprove and dismiss some of our most cherished truths.

The process of discovery is of course a central ingredient of virtually every scientific discipline. But true discovery takes many forms. Perhaps the most profound experience of all is the discovery that one has the capacity to directly and qualitatively affect the lives of other human beings, even of faraway strangers—for better or for worse—through one's work. I hope all of those associated with this book will have discovered this anew.

WILLIAM H. SHORE

WALKING IN TIME

TIMOTHY FERRIS

I'm walkin'
Yes indeed.
I'm talkin'
About you and me . . .
                    —*Fats Domino*

So, little by little, time brings out each several thing into view,
and reason raises it up into the shores of light.
                    —*Lucretius*

Once, years ago when young and proud, I tried by walking for a single day to grasp the enormity of the past.

Space I knew a bit about. Since childhood I'd studied astronomy books and the boundless book of the night sky, acquiring a sense of cosmic cartography. I knew the directions and distances of scores of stars, was familiar with the reefs and shoals of the Milky Way thousands of light-years out, and could find my way down lanky archipelagoes of galaxies stretching from here to the core of the Virgo cluster. I felt at home in the depths of the sky.

Time, however, baffled me. I'd labored to memorize when the Triassic gave way to the Jurassic and whether the Frasnian belonged to the Devonian or the Silurian period, but had failed to establish a navigator's claim on the dark waters of the past. Time gone by remained a trackless wasteland to me, negotiable by map but not by intuition. Now, in my early thirties, I had in mind to remedy my bewilderment through a physical act, a kind of walkabout. I would attempt to walk backward in time, imagining that each step transported me one thousand years into the past.

I decided to start in Times Square. (I was living in Manhattan in those days.) My plan was to walk up Broadway, the oldest street in New York City, then across the George Washington Bridge, through the dinosaur-haunted New Jersey plains, and south to Arlington, New Jersey. There I would visit my father's grave—he'd died the previous year—and look in on his brother Bill, who'd just suffered a stroke and was hospitalized nearby in Kearny, the town where my grandparents, immigrants from Belfast, Ireland, had raised their five sons.

Times Square at dawn that clear May morning was colorless as a charcoal sketch. Fish-gray buildings rose up from the black asphalt streets into an off-white sky. A bone-white shaft of sunshine cut across Manhattan from the East River, backlighting whirling white steam devils that hovered over loose manhole covers, etching long black shadows behind dancing sheets of discarded newsprint. It seemed a good day to depart from the present. I stationed myself on the northwest corner of Broadway and Forty-second Street, faced north, squared my shoulders, took a deep breath, and headed upstream in time.

A single step, and my imagination was transported back one thousand years, to A.D. 978. The city vanished. Manhattan, the "Island of Hills," was cloaked in virgin timber. Broadway alone persisted, a foot-wide Indian trail winding through the trees, one of a network of long-distance avenues comprising a Native American rapid-transit system. An able messenger, moving swiftly and silently on moccasined feet, could run the 150 miles from here to Albany with but a single rest stop, that at a spring on what would become the Albany Post Road near modern Poughkeepsie.

The Unami Lenape lived in these parts when the first settlers arrived from Europe. They were immigrants themselves: They'd made their way here in about the fourteenth century A.D. from the shores of a vast, frigid body of tidal water far to the northwest—one of the Great Lakes, I assume, or perhaps Hudson Bay. "It freezes where they lived," reads a tribal pictograph describing their journey. "It snows, it storms, it is cold where they lived. . . . All said to their priest, 'LET US GO.'" So they headed southeast and didn't stop until they reached another body of tidewater, which happened to be the Hudson. Here they hunted *mides, tomaque,* and *lingwes* (elk, beaver, and wildcat), caught *lamiss* (fish), and called the winter snow *whinne,* their word for "beautiful."

Meanwhile, Viking explorers were coasting Greenland; they ruled Ireland, where in thirty generations my father would be born. I took a second step, and the woods grew slightly colder. The year was 22 B.C. A more primitive tribe that predated the Lenape lived here then. Some may have occupied a cave dwelling in a glen at the northernmost tip of Manhattan, part of what is now Inwood Hill Park; certainly they had a settlement farther south, at the foot of Dyckman Street. Far away, in the world of the written word, Chinese junks were running porcelain to India and Virgil was drafting the *Aeneid,* the poem he'd always wanted to write but would not live to complete.

A third step took me into the eleventh century B.C. Olmec artisans were carving stone in Mexico. In Babylon the Chaldeans were learning to measure time by water clock. Four steps, and I was upstream of Hammurabi; five steps, of the Sumerian Gilgamesh epic; at six steps I had reached a time prior to the invention of writing. By step twelve the Hudson Valley was devoid of human occupants, and these hills belonged to the wolves, deer, bear, geese, and eagles.

At this point a shining finger of ice was busily refilling the fjord we call the Hudson River, burying Manhattan beneath a glacier a thousand feet thick. It took me about a hundred steps to get through this, the most recent ice age. No humans are known to have reached North America that early. A lot was going on in Europe, however. The brain size of genus *Homo* swelled dramatically during the Ice Ages, culminating during the last glaciation in an explosion of human invention that produced bricks and mortar, musical instruments, jewelry, the bow and arrow, the sewing needle, the domestication of animals, the cultivation of potatoes, pumpkins, and beans, and the inscription of maps and astronomical records on bone. Such a flurry of activity suggests that the Ice Ages somehow sparked the rise of intelligence, but nobody knows just how this might have happened.

Neanderthals were around at that time, their bodies *and* brains a fifth larger than those of *Homo sapiens,* yet for some reason the Neanderthals perished and the *Homo sapiens* didn't. Nobody knows why that was, either.

I passed the dawn of genus *Homo* somewhere around Columbus Circle, two million years ago. Meanwhile, present-day Manhattan was beginning to wake up. A breathless young man towing a pup on a leash borrowed my pen to scrawl, on a white paper bag retrieved from a trash can, the name and telephone number of a young lady he'd just met on the street. A block farther along, I encountered a derelict who was staggering down the sidewalk, muttering furiously to himself, "God may have said it, but I said it *first!*"

I strolled up Central Park West, listening to birds singing in trees cool with the mint-fresh oxygen of early spring, keeping an eye out for signs of ancient times. Sidney Horenstein, a geologist from the Bronx who treats cities like field digs, had taught me to be alert for urban limestone. Inspecting the facade of the Universalist church at Seventy-sixth Street, I could indeed make out the tiny, bleached-white skeletons of marine creatures—the spiraling *Straparolus,* fanlike *Spirifer,* and *Archimedes,* a perfect replica of an Archimedes's screw—that some 330 million years ago inhabited a shallow sea in what is now Indiana, where this stone facing had been mined. Just inside the park at Seventy-seventh Street I spotted a ginkgo tree, emissary of a survivalist species that dates back 200 million years, having endured even the calamity that killed off the dinosaurs. Here and there in the park I could see ragged, mica-flecked outcroppings of bedrock that congealed from mud 450 million years ago.

I detoured through the American Museum of Natural History to view, in glass cases, models of minute marine life forms that thrived 65 to 300 million years ago—ammonoids shaped like kazoos, horned nautiloids richly adorned as a monarch's chariot, and rugose corals gaudy as Las Vegas by night. All these temporal totems were beyond my reach; if

I walked twenty miles today, as planned, I'd get back only some 40 million years.

At Eighty-second Street I crossed to Riverside Park, animated now by bicyclers and joggers, then ambled north toward Washington Heights and the George Washington Bridge. As the sun climbed high in the sky, the temperature rose to nearly ninety degrees. In Harlem, children were playing in the spray of fire hydrants.

To the eye of my imagination, the world looked very different. With time moving at a thousand years a step, day and night blurred together like the frames of a motion picture. If I paused to examine the night sky, I could recognize nothing: the constellations, scrambled by the wanderings of the individual stars in their orbits round the center of the galaxy, had been altered beyond recognition since I'd crossed Forty-fifth Street. By day the land bowed beneath the weight of successive ice sheets, then rose again as the ice receded, reclaiming sunken shores from the seas. I thought of Tennyson:

> There rolls the deep where grew the tree.
>   O earth, what changes hast thou seen!
>   There where the long street roars hath been
> The stillness of the central sea.
>
> The hills are shadows, and they flow
>   From form to form, and nothing stands;
>   They melt like mist, the solid lands,
> Like clouds they shape themselves and go.

At the midpoint of the George Washington bridge I paused to look out over the sparkling Hudson waters. The bridge cables ticked and swayed with the fitful stresses of passing traffic. To the right stood the Palisades, to the left the Manhattan skyline. We think of the Palisades as natural and permanent, the city as artificial and transitory; and the Manhattan towers do indeed seem ephemeral as a soap bubble if viewed from the perspective of an accelerated time scale. Yet both are but passing forms assumed by matter, which is but congealed energy, and both will fall in time.

Though I was back a long way now—over 17 million years—much of the life on earth would still have seemed familiar: there were apes and monkeys then, and deer, giraffes, pigs, whales and dolphins, and songbirds (though no elephants yet, or dogs or baboons). The planet, too, probably looked much as it does today, if we disregard such details as pulsating ice caps and the submersion of Iceland.

The sun was near the zenith. I was beginning to appreciate how little progress I would make in one day's walk.

Just past the New Jersey entrance to the bridge I came upon two men who were struggling to repair a truck in the shade of an overpass. A hand-lettered sign, inscribed in spidery blue ink on a sheet of paper taped to the side of the truck, read, French Quiche—Homemade. The quiche was spoiling in the back of the truck. The men were sweating. They looked discouraged. The truck must have broken down hours earlier, frustrating their hopes of making a few dollars on a Saturday morning.

I knew how they felt. The sight reminded me of a day when I was nine years old. My father, down on his luck, was making a living working odd jobs by day and writing short stories by night. He spent the last of his savings on a load of cement bags. He loaded the cement on a borrowed flatbed truck and drove it to a construction site deep in the Florida Everglades. On the way the truck broke down, and while he was fixing it a thunderstorm came in from the northwest and drenched the cement. By the time he arrived at the construction site it had been abandoned for the night. He unloaded what he was certain was a ruined batch of soggy cement, made the long drive home, then collapsed on the bed, his coveralls bleached white with cement dust, too exhausted and discouraged to eat or bathe or say a word. Three days later a check arrived in the mail. When my father called to see if there had been some mistake, the construction foreman reassured him that bagged cement gives off so much heat that rain can't harm it. It's full of limestone, after all—the skeletons of ancient sea creatures, organic matter as potent in its way as a humus pile. We bought groceries that day, and soon thereafter a magazine bought a story my father had sent them, and things began to look up again.

I strolled through the Revolutionary War community of Fort Lee, favoring pools of leafy shade to escape the midday heat. The American Revolution had occurred only 2.5 inches into my first step. Twenty million years now lay behind. Yet the age of the dinosaurs, who flourished in these parts, was still 45 million years ahead. Their remains, buried hundreds of feet underground, were more removed from my imagined point in time than were the passing autos and buses of the present.

I turned south and hiked through a waste of despoiled grasslands east of the New Jersey Turnpike. (The origin of grass lay in this part of the past, some 24 million years ago.) Sequoias once grew in New Jersey, before colder climates killed them off everywhere but California. My feet were tired and I was lost in time, unable now consistently to distinguish one millennium from another, but when two cheerful young women in a bright yellow convertible stopped and offered me a ride I turned them down, determined to finish my trek. I accepted a drink of lemonade from their thermos bottle, though, and we chatted by the side of the road about next year's prospects for the New York Giants—a subject of interest to the driver, who said she was in the process of divorcing a Giants lineman. With a wave good-bye, they roared off across the flats of the dinosaur burial ground. I trudged on into the gloom of the Eocene–Oligocene turnover, 35 million years ago, when a comet is said to have struck the earth and set the world ablaze, dooming enough species to make way for the rise of rabbits, squirrels, gophers, and dogs.

It was late afternoon when I reached my father's grave. The air was cool and fresh there, in a grove of old elms. I read the epitaph—Thomas Addis Ferris, 1908–1977, Beloved Husband and Father—then sat back against the tombstone and looked out across hillsides set with hundreds more tombstones that jutted up at subtle angles, obdurate as old teeth. Indians are buried hereabouts, too, I remembered, their bodies curled up like fetuses in shallow, winter-hewn graves lined with sweet grass.

Not far from where I sat had been the tennis courts where my dad as a boy had rolled the clay at dawn in return for morning court time, the diamond where he'd pitched for the local baseball team, the seaside camp where he'd coached boxing, the docks from which he'd shipped out aboard freighters bound for Rio and San Francisco. He was three years old when he was brought to Kearny by my grandfather, a machinist trained on the lathes of Harlen and Wolf's shipyards in Belfast. Once when I asked my grandfather what he knew about his ancestry he said that *his* grandfather had been a gameskeeper on an estate in the north of Ireland. Of the family prior to that he knew nothing.

I reckoned that my walk had taken me 44 million years back in time, and of that I knew next to nothing, too. I'd translated into sore feet and aching calves what I'd learned from books about the evolution of horses and rhinos in the Eocene, of the emergence of the Bering land bridge and the separation of Australia from Antarctica, but in truth all had been foreshortened into something approaching chaos. I'd lost my battle with the past.

I took a cab from the cemetery to the hospital. There I found my uncle Bill, hooked up to tubes and wires in the intensive-care ward. The stroke had paralyzed the left side of his body, and he looked shocky and emaciated. But he greeted me with a lopsided grin, accepted a harmonica I gave him, and soon was singing and playing a spirited rendition of "The Sinking of the Titanic." His three surviving brothers, James, Jack, and Stanley, arrived, and we all sang along until the nurses shooed us out. Then we drove over to Bill's house. We pulled up in front while Uncle Jim was recounting a particularly funny tale about a boxing match Jack and Bill fought as teenagers, so we all stayed in the car while he finished.

"Who won that fight, Jack?" Jim asked.

"Oh, Bill beat me," Jack said modestly. Jack was unfailingly modest, a habit for which his brothers seldom forgave him.

"That's what Bill told me, too," said Jim.

Bill's wife, my aunt Dot, a diminutive woman with a powerful voice, emerged from the front door of the house and scowled at the car. The dying rays of the setting sun framed her aproned figure with a corona of gold. It was one of those sights that you know instantly you will never forget.

"What are you doing out *there?*" Dot shouted, in a peal that could have carried to the next valley across the rolling hills of Ireland. "Don't keep the good stories *outside,* boys. Bring the good stories in *here.*" And in we went, amid ringing laughter, just before dark.

FIRST BIRTH

ALAN LIGHTMAN

An obstetrician friend of mine told me that she never gets tired of delivering babies. She's been at it for twenty years now, sometimes bringing into the universe two or three babies in one day, and each birth still excites her. "It's a mystery," she says. "First there is nothing, then there is something—a life, a personality."

Very recently, physicists and astronomers have made new progress in understanding the birth of the universe. The progress has come from equations and inference—no one was there to watch the delivery—but it is progress nonetheless, in the way that modern scientists often advance in territory beyond everyday experience. The birth of the universe would of course be the very first birth. And a critical question seems to be, Was there ever a time when there was *nothing?* No galaxies, no stars, no planets, no atoms, no electrons, no space, no time. Nothing.

The leading theory of the origin and evolution of the universe, the Big Bang theory, does not actually address the origin, the birth. The term *Big Bang* was coined by the British physicist Fred Hoyle on the BBC in 1950, but the theory itself dates back to the 1920s. In 1922, Alexander Friedmann, a Russian meteorologist and mathematician, proposed a detailed mathematical theory for the evolution of the cosmos according to which the universe began in a state of extremely high density and has been expanding and thinning out ever since. The explosive beginning at high density was the Big Bang. Friedmann's work was all pencil and paper. There was no evidence that the universe was expanding or doing anything else. In fact, scientists debated what the universe was. It was known that the visible stars in the sky were distributed in a disklike shape called the *Milky Way*. But it was not known whether certain misty patches in the sky, puzzled over for centuries, were small collections of stars inside the Milky Way or themselves large galaxies of stars far outside the confines of the Milky Way.

In the middle to late 1920s, the American astronomer Edwin Hubble measured the distances from Earth to some of the misty patches and found that many of them indeed inhabit remote regions of space well beyond the Milky Way. Thus the universe is not a single island of stars in a vast empty sea but an archipelago of many such islands scattered about on a grand scale. The distances are almost impossible to fathom. The circumference of the Earth is about 24,000 miles, the distance from Earth to the Moon about 250,000 miles, to the Sun about 100 million miles. The distance to the star nearest the Sun, Alpha Centauri, is about 25 trillion (25,000,000,000,000) miles. To measure greater distances, it is convenient to use the light-year, which represents the distance light travels in a year—about 6 trillion miles. In these terms, Alpha Centauri is about 4 light-years away. Our galaxy, the Milky Way, a congregation of about 100 billion stars orbiting about each other under their mutual gravity, is 100,000 light-years in diameter. In other words, it takes a light ray 100,000 years to cross from one side of the Milky Way to the other. The galaxy nearest to us, Andromeda, is about 2 million light-years away. Our telescopes have identified cosmic objects 10 billion light-years away.

Hubble discovered something else. The other galaxies are racing outward from us in all directions, with speeds proportional to their distances. This result was just as expected from the mathematical theory of Friedmann and others, and it gave observational evidence that space is not the rigid system of fixed stars described by the poets, but instead a dynamic, expanding membrane on which the galaxies are carried apart from each other like ink marks on a stretching rubber band.

If the galaxies are flying away from each other, then they must have been closer together in the past. By measuring the rate at which the cosmic rubber band is stretching, or more exactly, the rate at which galaxies are moving outward, Hubble and others estimated how long ago it was that all of the galaxies were crammed on top of each other—about 10 billion years. That moment would seem to have been the beginning of the universe. In fact, a universal age of 10 billion years is in rough agreement with the age of the Earth as obtained by radioactive dating of uranium ore and with the ages of clusters of stars as obtained by their evolution and change.

Since the 1960s, detailed calculations using Friedmann's equations and tested laws of physics have traced the history of the cosmos back to the first millionth of a second after the beginning. The temperatures at this epoch were so hot (10 trillion degrees centigrade) that even individual atoms could not hold together, much less whole stars or planets. The entire material content of the universe—all that would later form atoms and molecules and planets and people—would have existed in the pristine form of subatomic particles, careening through space at great speed.

But one-millionth of a second after the beginning is not the beginning. The problem with attempting to push Friedmann's theory back to the very beginning, to the birth of the universe, is that at a certain point the theory falls apart. Friedmann's theory of an expanding cosmos was based on Albert Einstein's theory of gravity, called *general relativity,* developed in 1915. Einstein's theory, in turn, was designed to describe the world on the scale of planets, stars, and galaxies. It was not designed to describe the subatomic world. But sufficiently close to the beginning, the material contents of the cosmos would have been so densely crammed together that the entire universe would have behaved like a subatomic particle. There is indeed a branch of physics, called *quantum physics,* that accurately describes the subatomic world, but quantum physics has not yet been successfully combined with Einstein's theory of gravity. (One recent attempt at a combination, called *superstring theory,* is not yet manageable mathematically and remains experimentally unconfirmed.) Thus, there is one theory that works well in the large, general relativity, and one that works well in the small, quantum physics, but no good theory for what happens when the large becomes small. And physicists, as if knowing how to navigate either in the dark or in the rain, but not in both, now find that they must venture forth on a rainy night to understand the very first moments, the *quantum era,* of the universe. We can estimate when the quantum era occurred—the first 0.0000000000000000000000000-0000000000000000001 (one ten millionth trillionth trillionth trillionth) seconds after the beginning—but we cannot reliably calculate what happened during that time. The quantum era is a thin, dark mist that shrouds and surrounds whatever happened at the beginning.

Nevertheless, physicists have attempted to see through this mist. One such attempt, by Alex Vilenkin of Tufts University, is described in a paper with the interesting title "Creation of Universes from Nothing." In brief, Vilenkin—a quiet, modest fellow in person—proposes that before there was something, there was nothing. His idea follows from a strange property of the subatomic world that allows matter to appear out of nothing, provided it disappears again with sufficient haste. The sudden appearance and disappearance of subatomic particles leads to small but real effects in the energy levels of atoms, which remarkably have been confirmed in the laboratory. The nothing out of which electrons and quarks and other subatomic particles may materialize is quaintly called the *vacuum.* Now, if an electron can pop up out of nowhere, then why not a whole universe, Vilenkin asks. After all, in the quantum era the entire universe behaved like a subatomic particle.

One might worry, if Vilenkin is right, why the whole universe doesn't disappear again quickly, sneaking back into that nothingness from which it emerged. The resolution of this conundrum might be, in part, that the universe was created with net-zero energy. Gravitational energy is negative, and there may have been sufficient negative gravitational energy in the infant universe to counterbalance the positive energy that later formed matter and stars. By contrast, when subatomic particles appear out of nothing in the lab, they are created with hardly any gravitational energy to counterbalance the positive energy of their mass, and so they must soon disappear to avoid a permanent creation of net-positive energy. The universe as a whole, with vast stores of negative gravitational energy, may not have had to repay any energy deficit.

A more vexing question is whether Vilenkin's nothing truly is nothing. According to the theory of quantum physics, the physicists' vacuum is jammed full with information about all the various types of particles that it might suddenly spit forth. For example, only certain types of subatomic particles with certain masses can exist, and the vacuum must somehow know what it is allowed to materialize and what not. Indeed, in some interpretations of the quantum theory, the multitudes of subatomic particles that occupy our world already have ghostlike existences in the vacuum, waiting to be summoned forth by random microscopic events. In this picture, Vilenkin's nothing would actually be a haze of potential universes waiting to come into being. The potential universes would have different properties—for example, different values for the mass of the electron or the speed of light, perhaps six dimensions of space instead of three, perhaps a different kind of time—and our universe would have just happened to be one of many possibilities. It is difficult to conceive of this blur of potential universes. Did space exist? Did time exist? Almost certainly not in any form we would recognize. If the answer to "Was there ever a time when there was nothing?" is "Yes," we might then follow with "What is nothing?"

Another recent effort to confront the beginning has been undertaken by Andrei Linde, formerly of the Lebedev Physical Institute in Moscow and now at Stanford. In person, Linde is the opposite of Vilenkin. He has a flair for the dramatic. He says things intended to provoke his listeners, and he illustrates his lectures with racy cartoons drawn by himself. Linde is one of the architects of the *inflationary universe model,* a modification of the Big Bang theory. According to this model, which has plenty of mathematics to back it up, the vacuum of the infant universe may have temporarily behaved as if it had negative gravity. Negative gravity repels rather than attracts, and such a state of affairs would have caused the universe to expand briefly at an enormously rapid rate. While the universe was still much less than a second old, gravity would have converted back to its regular, attractive form, the inflationary epoch of wildly fast expansion would have been over, and the universe would have returned to the leisurely rate of expansion of the standard Big Bang theory.

Linde speculates that under certain conditions, an inflating universe can separate into different pieces completely cut off from each other—

in effect, different universes. Each of the new pieces can repeat the process in a random way, with each universe spawning many new universes. Individual universes might expand, contract, and then crush themselves out of existence. Individual universes might come and go, like the fleeting worlds of the ancient Greek philosopher Anaximander, but the collection of universes never dies. New universes are always being born. According to Linde, if we traced the history of our own universe back to the quantum era, we would find an umbilical connection to a parent universe, and that parent to a parent and so on, back into the infinity of time. Thus, Linde says, before there was something, there was another something. There was never a time when there was nothing.

Linde's reproducing universes have some similarity to the steady-state model, proposed by Fred Hoyle and others in the 1940s. According to this model, there was no beginning. The universe has always existed in pretty much the same state as it does now. The movement of galaxies away from us is only a "local" phenomenon extending over 10 billion light-years or so. Hoyle's current thinking is that there may have been many little "big bangs," on local scales, but that overall the universe is steady and eternal.

We might occasionally remind ourselves that Vilenkin's and Linde's ideas about the beginning are highly theoretical, like Friedmann's scribblings seventy years ago. Observational evidence in cosmology is scant, and very little of it relates to the infant universe. The tiny unevenness in the cosmic radio waves discovered last spring by George Smoot of Berkeley and his collaborators is a vital clue to our understanding of how the gaseous material filling the universe could have gathered itself into galaxies, but that discovery does not tell us whence came the material itself. In other terms, the cosmic radio waves that we observe with our radio telescopes, the most ancient light that we see, was produced when the cosmos was already 300,000 years old.

According to the *Enuma Elish*, the ancient Babylonian story of Creation, the world began in liquid chaos. In time, the slow seep of silt created the gods Lahmu and Lahamu, who together stretched into a giant ring to form the horizon. From the upper side of the ring grew the heavens, and from the lower grew the earth.

Mesopotamia is a region built by silt at the juncture of the Tigris and Euphrates, and this watery world provided the images found in the *Enuma Elish*. The Big Bang theory, extended by the speculations of Vilenkin, Linde, and others, is our modern *Enuma Elish*. In place of the Tigris and Euphrates, we have quantum physics and general relativity.

In a thousand years, quantum physics and general relativity may be part of a unified theory, or they may be replaced altogether by some new version of the beginning. Yet it seems likely that the beginning will remain a mystery. For a universal feature of knowledge is that one must get outside of a thing to understand it. To understand the crucial aspect of an airplane, one must stand on the ground and watch it move through the air overhead; an understanding of the essentials of American society required the visit of Alexis de Tocqueville from France; a famous theorem of mathematics, Gödel's theorem, says that the truth or falsity of certain statements in each branch of mathematics cannot be ascertained without going beyond that branch to another area of mathematics. The universe, by definition, is everything there is. How do we get outside of it? We have only one universe, and all of our conceptions of time and space, all of our laws of nature, are based on this one universe. How can we conceive of a platform from which to witness our universe come into being?

In each age, scientists have thought they were on the verge of discovering all the laws of nature. But the birth of the universe, thankfully, will remain cloaked in mystery.

Somebody Else's Cosmology

Anthony Aveni

Our cosmology is a contemporary set of beliefs about how the universe is structured, where it came from, and what will happen to it. When we say we believe in the Big Bang we mean that we are convinced—at least today—that our universe had a definite beginning point in time and that the beginning was violent—an awesome flash that happened everywhere, followed by a rapid expansion and a drop in temperature that have been occurring ever since.

What gives us the unshakable faith that this story is the only true account of where we came from? Answer: our equally unshakable faith in a process of reasoning and experimentation, a quantitative way of knowing that we call science.

Science works through the creation of ideas that can be tested in the real world. Through the results of these tests—we call them *experiments*—we modify our ideas, adjusting or replacing their defective component parts, in order to build even better ideas. For example, half a millennium ago the Polish astronomer Nicolaus Copernicus figured that if one imagined the Earth to be one among a set of planets circling the Sun, it would be possible to explain the periodic backward motion of other planets against the background of the stars in a more elegant and efficient way than if one assumed the Earth were fixed and the planets did all the moving on their own. His idea of a Sun-fixed rather than Earth-fixed planetary system has been good enough to last for over five hundred years. The key words to describe his idea are *elegant* and *efficient*. What is elegant is all a matter of subjective taste, and not every cosmological idea needs to be efficient. Copernicus's idea suits us because it meets our criteria for judging what a true story ought to be like.

For the scientist, a good idea is one that precisely accounts for what we see in nature, down to the minutest detail. Being able to predict in advance the outcome of a close observation of one of nature's myriad phenomena is central to this way of knowing nature that we call *science*.

But there is another pillar upon which all of science rests. It is the belief that all of the universe and its parts exist and operate quite independently of any human concern. As the fashionable cosmologist would say, the universe is observer independent. The business of science rests upon the ability to create testable scientific themes in an objective universe, one that functions strictly for its own sake.

Now, not all of these truths are self-evident. They are objective only if we shut our eyes and ears to eons of human history and pay no attention to what wise people from other civilizations in other times and places have had to say about the structure of the world around them. Our truths have become established because of a kind of evolution, not Darwinian evolution, but the evolution of culture—the slow buildup of ideas and observations that form what the anthropologist would call a *cultural norm,* a set of beliefs we call *common sense*. In my view, science is but one of an infinity of possible common senses regarding the natural world. It is the one that works for us.

I suspect that there are as many ways of knowing nature and as many cosmologies as there are (and were) human cultures in the world to dream them up. What follows is one example—the dream of an ancient people who inhabited the same planet and saw the same sky as we.

Have a look at the drawing on page 19. It is a leaf out of a folded-screen book from ancient Mexico—one of the few that survived destruction by the Spanish conquistadors. The design—I call it a *cosmogram*—is a map that shows the structure of the universe. Its outline takes the form of a Maltese cross merged with a floral cross; it is drawn on deer hide, lime coated and colorfully decorated with mineral dyes.

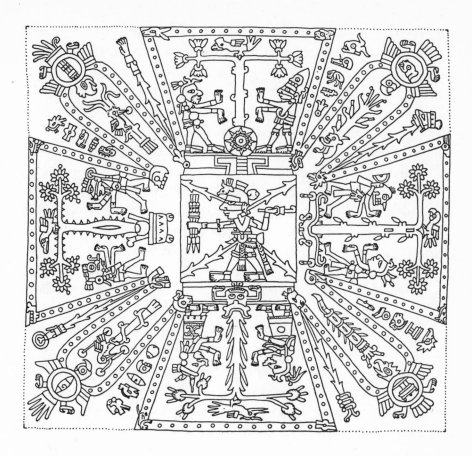

The cross shows at the center not the Earth, not the Sun, not even the Milky Way Galaxy, but a creator god. And this immediately demonstrates that the ancient Mesoamerican conception of the universe is quite unlike ours. The presence of a god at the ideological (rather than geometrical) center implies that their world does not operate for its own sake but instead that their god—or more accurately, gods—have some input. In this universe the gods decide the course of natural events, at least in part, so that scientific experiments do not have predictable outcomes that can be used to validate or refine scientific theories.

Now look at the blood that flows in streams toward the Creator god, one from each of the four cosmic directions. In addition to active deities, the old Mesoamerican cosmology has something else that ours doesn't have—purpose. Their universe had a moral component. People had a role to play, and blood was the key. Those Mesoamericans believed that through the shedding of human blood—sacrificial blood—the gods could be paid back for their gifts of rain, bountiful crops, and human productivity. Clearly, this cosmological tale is a reciprocal one; it stresses the joining of people and gods in a kind of communal social relationship, working together in harmony.

The philosophy goes something like this: We work for the gods, who are superior but largely benevolent beings who control the universe. We are not all-powerful, and we understand that we can never fully master the mysteries of the universe. But we are confident that we were put here to help keep things in working order. The blood we offer our gods guarantees that the Sun will forever remain on his course and continue to nurture us, that he will never be overcome by the forces of darkness and chaos. This is our purpose: to serve the gods who created both humanity and the rest of the universe.

Follow the bloodstreams out to the four corners of the universe and you will find parts of the human body attached to them—here a head, there a hand or foot. The idea of associating body parts with celestial domains may be familiar to those acquainted with medieval astrology, according to which Gemini governs the loins while Pisces is in charge of the feet. This associative way of thinking is very different from our idea of cause and effect. Its logic is metaphorical, like that which characterizes adolescence as the spring of life and old age as the winter.

The Sun, for example, is associated with the east, where it first appears, bringing light and life into existence. East is at the top in the diagram, where the rayed solar disk rests on an altar that symbolizes the sacred place where the Sun is worshiped. In the tropics this altar most probably would have been positioned in an open plaza in front of one of the temples dedicated to the cult of the Sun god. Religious Europeans and Americans who are used to worshiping in great churches and cathedrals often have difficulty appreciating the way these outdoor environments were specifically tailored to nature worship. Sometimes holy temples were specifically oriented to the Sun god's approach so that at the correct time and place he could be given his ritual offerings. The eastern flaps of the Maltese cross that frames the Sun are outlined in red, which surely suggests the color of the dawn. And the tips of the elongated flower petals to either side likely represent the seasonal migration of the Sun across the horizon, from winter, when it lies south of east, to summer, when it is positioned north of east.

The trees that lie in each of the four directions and the birds perched in their limbs have not all been identified by modern researchers. Those we can identify do seem to have a logical association with a direction from central Mexico. For example, the northerly directional tree is framed by a yellow sky and could be a saguaro cactus, which flourishes in northern Mexico and the extreme southwestern U.S.

Whoever drew our ancient Mesoamerican cosmogram seems to have been trying to assign a place to everything and to put everything in its place. The diagram's goal is all-embracing, while that of our Big Bang cosmology is particular. Our cosmic story has no place for trees and birds because these are not thought to be part of the first steps in Creation, which we imagine as a time-dependent process of ever-increasing complexity. We have a separate Creation theory to account for plants and animals—one framed not primarily by physics but by zoology and molecular biology. Likewise, there is no place for color directions in our cosmology.

Rather than addressing or tackling the when and how of Creation, the Mesoamerican cosmogram tries to envelop and interconnect all aspects of space and time. Time is represented by the dots that decorate the periphery of the Maltese cross. As on some kind of cosmic Parcheesi board, the dots go all the way around the diagram—they literally envelop everything in the physical universe. Count the dots and you will arrive at the number of days in the cycle of time these people used to measure out their existence. This was not the 365-day solar year (though there *is* a way to measure that in the cosmogram, too) but the 260-day cycle, a period made up of 20 named days of a week, each of which recurred 13 times in a cycle, just as we have 52 Sundays in a year.

There is currently a great debate among scholars of Mesoamerican history about the origin of this unique cycle. (Oddly enough, we find it only among the civilizations of ancient Mexico.)

Two hundred sixty is 13 times 20, and 20 is the number of human fingers and toes. When we developed systems of notation, we could logically have predicted they would be either decisional (based on 10, as ours is) or vigesimal (based on 20)—which the Mesoamerican one was. Thirteen was the number of layers of heaven in their upper world. But 260 was also the approximate number of days in the agricultural season in the southern Guatemalan highlands where this calendar likely originated. In fact, indigenous people there still refer to this time cycle as the "corn year." Also, native women still tabulate the interval between conception and birth as 260 days. Moreover, the mean morning- and evening-star appearance intervals of the planet we call Venus—Quetzalcoatl to the people of the highlands and Kukulcan to the Maya of Yucatan, named after a major male god in the ancient Mesoamerican pantheon—are close to 260 days in length. The eclipse half year (173.5 days), a basic period that governed a calendar keeper's ability to predict eclipses, coincided with the 260-day cycle in the ratio of 2 to 3. This means that if we make use of the day numbers on an imaginary 260-day time wheel to generate warnings of eclipses, the same general set of numbers—they would be located ⅔ of a revolution apart on the wheel—would be assignable to eclipses that actually can be seen to have occurred. Still another possible astronomical explanation for the origin of the 260-day cycle comes from the fact that in the southern regions of Mexico and Central America, the day when the Sun crosses the zenith (or overhead point) of the sky splits the 365-day seasonal year into unequal intervals of 260 and 105 days; that is, once the noonday Sun crosses the zenith (May 18) it remains in the northern part of the sky for the shorter interval; after it passes overhead a second time (July 26), it remains in the south for the duration of the year. This would have been a logical, highly visible way of marking the seasons in a two-season year.

Perhaps, just as modern scientists have discovered unifying concepts like angular momentum and the constant of gravitation by noticing that they are manifested in a host of different ways, someone in the first millennium B.C. realized that the gestation cycles of humans and maize and the cycles of certain heavenly bodies could all be unified in a single number—260.

Regardless of time's origin, in ancient Mesoamerican cosmology it envelops all that exists in space, and that is why the artist, like some Einstein from another culture, has seen fit to bind time and space together in this way. But what Einstein sought to achieve with his abstract equations, our unknown ancient artist-cosmologist has attempted to represent in a single coherent picture.

Who can say whether a society is correct or incorrect about the cosmology it creates? Our ideas and those of the ancient Mesoamericans serve different aims, goals, and purposes. Our criteria are different from theirs; our ways of getting at the truth—our very definitions of what we mean by *truth*—are incompatible with theirs.

What our three- or four-dimensional Big Bang universe and the painted cosmogram from preconquest Mexico have in common is that each tries to tell a story about the nature of the universe. Each story is fundamentally about finding an order that we are all confident must already be there—waiting for us to unveil or unmask it. While we moderns have committed ourselves to a strategy of looking at the abstract mechanisms we believe underlie what we superficially observe—while we strip down, separate, dissect, and reintegrate the stuff of the universe according to our own set of rules, confident that it has no spirit, no soul—the ancient Mesoamericans animated, associated, and found direct and immediate moral purpose in everything they saw and sensed. Different time, different place, different story.

A Detective Story

BRUCE GREGORY

In a detective story, the protagonist examines the scene of the crime—footprints in a muddy driveway, a windowsill free of fingerprints, an opened letter on a desk. In the detective's mind, these apparently unrelated and possibly trivial facts become the essential elements in a carefully woven story pointing inevitably toward the guilty party. But is the story true? The people involved are gathered together. The detective reveals his impressive reasoning. The guilty party blurts out a confession, and the detective's logic is vindicated.

Like detectives, scientists build stories around often fragmentary evidence. The story astrophysicists tell about the origin of the universe is based on just such evidence. Astrophysicists tell us that the universe began some 10 or 20 billion years ago as an unimaginably intense burst of energy called by almost everyone the *Big Bang*. (Lewis Thomas describes the Big Bang more accurately, and more poetically, as the "great light." Some readers may recall the words of Genesis: "And God said, 'let there be light,' and there was light.") How did astrophysicists come to invent this story? On what evidence is it based? Why should we believe it?

Unfortunately, the universe, unlike the criminal in a detective story, does not obligingly "confess." The universe sometimes lets scientists know that their theories are wrong, but it never assures them that any theory is right. Rather, scientists must be content with a version of the Scottish verdict "not proven"—in science the verdict is always "not yet disproven," for new evidence might be uncovered that would upset scientists' most cherished beliefs.

With any explanation in a mystery story, it is not only important that the detective's version be plausible, but equally essential that there be no other plausible explanations. In general, astronomers base their story of the origin of the universe on three fundamental pieces of evidence. Before we accept the astronomers' story, however, we should act as a jury and examine each of these pieces of evidence to see if there are no plausible alternatives—if the story really holds together "beyond a reasonable doubt."

What kind of evidence have astronomers gathered that would bear on a question as remote in time as the origin of the universe? It is important to keep in mind that all astronomers have to work with is the "light" from distant objects. (Some of this "light" is in the form of X rays and radio waves, but the principle is the same. Also, particles from distant objects occasionally arrive on the Earth, but the origins of these interlopers are often difficult to determine.)

Almost everything astronomers know about the distant universe is an inference based on studying the photons that make up the light that reaches the Earth from distant objects. Furthermore, photons carry very little information. The only information of importance for this story comes from analyzing the direction from which the photons come and the energy of individual photons.

It takes time for photons to reach us from distant objects; the photons astronomers detect today left their sources many years ago—in some cases, astronomers believe, billions of years ago. In this sense astronomers, unlike geologists, have a direct view of the past. Nevertheless, the past they can "see" is much less dramatic than the still-earlier past whose existence they infer.

First, the muddy footprints. These footprints were uncovered in the early part of this century, when the American astronomer Vesto Slipher was observing smudges of light then called spiral nebulae. On very dark autumn and winter nights when there is no Moon, the brightest spiral nebula is visible to the naked eye as a faint patch of light in the constellation Andromeda.

What these smudges of light were and how far away they might be had been debated for over a century. Slipher thought they might be solar systems in the process of formation. In testing this notion, he photographically recorded the energies of the photons that make up the light from the nebulae. He found that in the light from many nebulae, the photons apparently emitted by the element calcium were lower in energy than the energy of photons emitted by calcium in the laboratory. This shift in energy is called a *redshift* because light made up of photons of low energy appears redder than light made up of photons of high energy.

The phenomenon was familiar to astronomers and physicists. Whenever the distance between a source of photons and an observer was increasing, such an energy shift was observed. Furthermore, if these energy shifts were due to velocities, the velocities involved were the greatest that had ever been observed—hundreds of kilometers per second. Slipher was not sure what these extremely high velocities could mean. Since he did not know the distances to the spiral nebulae, it was difficult for him to interpret the significance of their velocities.

Edwin Hubble was a lawyer, schoolteacher, and basketball coach turned astronomer, best known because the Hubble space telescope is named in his honor. He interpreted and extended Slipher's findings. Hubble had the advantage of using what was then the largest telescope in the world, with a mirror that was over eight feet in diameter; he also had the advantage of appreciating the full importance of a discovery that allowed him to infer the distances to the nebulae.

How could Hubble possibly measure these distances? An automatic camera incorporates a simple device to determine how wide to make the lens opening and how long to keep the shutter open in order to expose the film properly. Suppose a similar device were used to measure the apparent brightness of a 100-watt bulb at some distance from us. If the apparent brightness of a second 100-watt bulb were then measured and the two brightnesses compared, we could determine how far away the second bulb was in terms of the distance to the first.

By studying a series of photographs of the Magellanic star clouds, which are visible from the southern hemisphere, Henrietta Leavitt and Harlow Shapley of Harvard inferred that certain stars do come in standard luminosities, the way light bulbs do. Furthermore, these *standard stars* signal both their presence and their "wattages" by getting brighter and dimmer—the longer this change in brightness takes, the higher the wattage of the star. Thus, by putting a "stopwatch" on a star, it is possible to determine its wattage. By comparing this wattage with the apparent brightness of the star, Leavitt and Shapley could determine the distance to the star. If the star were located in a distant collection of stars, they could determine the distance to that collection.

Indeed, Hubble used standard stars in the Andromeda nebula to determine the distance to the nebula. He also applied Leavitt and Shapley's technique to other nebulae. Hubble was then able to infer that the nebulae are located at much greater distances from the Earth than anything in the Milky Way—they are vast "island universes" comparable in size with the Milky Way itself. We now call them galaxies.

In thinking about the scale of the universe, it helps to have some landmarks in mind. When you go out on a dark night, all the stars you can see are located in our "local" galaxy, the Milky Way. (What we call the *Milky Way* is simply the cloud of distant stars that outlines the structure of this local family.) Astronomers estimate that the Milky Way contains over 100 billion stars that range in size from less than one-hundredth the mass of our Sun to approximately one hundred times the mass of our Sun. Now, 100 billion stars is not easy to imagine. Nor is it easy to imagine the average distance between the stars.

One way to get some idea of the size of the universe is to imagine that each star is roughly the size of a healthy pea. In this case, 100 billion stars would more or less fill a large basketball arena. If you wanted to build a scale model of the Milky Way with these peas, you would have to expand the arena until each pea was separated by roughly 100 miles. This expanded building would have sides almost 3 million miles long. A scale model of the Milky Way with peas representing stars would stretch twelve times farther than the distance from the Earth to our Moon. You can imagine, then, that the Milky Way is largely empty space.

Now, building a model of the entire observable universe on this scale is obviously impractical, so we return to our original arena filled with peas. Only, this time, instead of letting each pea represent a star, let each pea represent an entire galaxy composed of 100 billion stars. You might guess that once again we will wind up with an unwieldy model, but in this case something quite different happens. Instead of separating our peas by 100 miles, as we did when each pea represented a star, we have only to separate the peas by less than a foot. *Relative to their sizes,* galaxies are much more closely packed in the universe than stars are packed in a galaxy. If a pea represents a galaxy, all the galaxies that astronomers can see would fit into a box with sides a few miles long. In comparison with the size of the galaxies that make it up, the observable universe is not all that large. We can now proceed with our mystery.

In order to measure the distance to remote galaxies, Hubble compared the brightnesses of nearby galaxies with the brightnesses of galaxies in which Leavitt and Shapley's standard stars were too faint to identify. He made the simplifying assumption that the more distant galaxies are approximately equal in luminosity to the nearer galaxies in which he had identified standard stars. This assumption allowed Hubble to infer the distances to the more distant galaxies from their apparent brightnesses. When he compared the inferred distance to a galaxy with its observed redshift, Hubble was sure he could see a relationship—the farther away from us a galaxy is, the faster it seems to be receding from us. Furthermore, if a galaxy is twice as far away, it is moving twice as fast.

But why are all the galaxies leaving us behind? One possibility is that the galaxies are escaping from the region of an explosion. The faster a galaxy is moving after some time has elapsed, the farther away from the site of the explosion it will be. From a position near the center of this expanding cloud of galaxies, individual galaxies—more or less evenly scattered in all directions—will be observed to obey a Hubble-like law.

Hubble, however, rejected this story. He realized that galaxies far from the center of the imagined explosion would see a very different picture than the one we see from Earth. We see galaxies distributed more or less uniformly through space. Observers on those distant galaxies would see many fewer galaxies as they looked away from the center of the explosion than they would see looking toward that center. This would imply that the Earth was at a special position at the hub of creation. Hubble favored a more democratic view in which observers everywhere in the universe would see a similar picture as they looked out into space.

Copernicus displaced the Earth from the center of the Solar System, and Shapley used the discovery of standard stars to show that the Sun is not at the center of the Milky Way. Hubble was not about to allow the Milky Way to become enthroned at the center of the universe. This decision was not based on any evidence, for it is obviously impossible to know what the universe looks like from the vantage point of galaxies far from the Milky Way, but on the conviction that the Earth must not hold a privileged position in the universe. This belief, similar to Darwin's conviction that human beings must not hold a privileged position in the animal kingdom, led Hubble to favor what appears to be an even stranger story.

Imagine two peas, each representing a galaxy, moving apart. An observer on either pea would see photons emitted by the other pea as redshifted. Now, imagine that instead of moving apart, the peas are attached to a straight piece of rubber band and the rubber band is stretched. Again an observer on either pea would see photons emitted by the other pea as redshifted. In fact, if there were more than two peas, an observer on *any* pea would see photons from the others redshifted in such a way that the farther apart any two peas were, the more the photons from those peas would be redshifted. In other words, an observer on any pea would see the same relationship between the distances and velocities of other peas that Hubble found between the distances and redshifts of galaxies.

It is not too difficult to imagine that the peas represent galaxies, but what does the rubber band represent? In Hubble's view, the nature of the rubber band was explained by Einstein's general theory of relativity as applied to the universe as a whole by the Russian physicist Alexander

Friedmann and the French cleric Georges Lemaître. Friedmann and Lemaître independently realized that Einstein's general theory of relativity calls for a universe that cannot be motionless but must either expand or contract. Einstein, however, was not at first impressed with their solutions. He told Lemaître, "Your calculations are correct but your physics is abominable!"

Since the Milky Way gives no sign of expanding or contracting, Einstein believed that the universe must be static. Nevertheless, Hubble's observations finally convinced Einstein. According to Friedmann and Lemaître, the rubber band represents space itself. Space is stretching, carrying the galaxies farther and farther apart, and in the process is stretching photons—lowering their energies as they travel between the galaxies.

The idea that space itself is expanding taxed the imaginations of some physicists in the 1920s and 1930s, but it is now commonly accepted, if no easier to visualize. Here we have the first important piece of evidence: We seem to have found the shoes that left their prints in the mud—the redshifts of distant galaxies are produced by the expansion of space itself. But is this the only possible explanation?

Gravity can also shift the energy of photons. Photons emitted by a very massive object are redder than photons emitted by a less-massive object of the same size. Perhaps gravity is the source of the redshifts of distant galaxies. However, if the redshift of a distant galaxy is produced by the amount of matter in the galaxy, there is no reason why the more distant galaxies would have larger redshifts—unless all galaxies were more massive in the past. But what would have happened to this mass? How could it have escaped from the galaxies, and where would it be now? Moreover, redshifts produced by gravity imply very dense matter, and many of these redshifts are associated with very tenuous gas. For these reasons, astronomers have rejected a gravitational interpretation of the redshifts of galaxies.

Another interpretation was suggested by the astronomer Fritz Zwicky, among others. In this model, space is not expanding and the galaxies are not moving apart as fast as they appear to be, but something is happening to the photons as they travel across the vast reaches of space.

Somehow they are losing energy, and the farther they have to travel, the more energy they lose. This story lacks credibility among scientists, however, because no suggested mechanism by which photons might age has proved plausible.

A different kind of argument also supports the expansion interpretation. The universe *should* be either expanding or contracting. It requires considerable effort to imagine a situation in which the universe is *not* expanding or contracting—either gravity would have to pull the galaxies together, or the galaxies would have to be moving apart rapidly enough to avoid gravitational collapse. But the galaxies can't be sitting still, at least not for very long, for if they were they would soon begin to fall together. It may seem surprising that the galaxies are flying apart, but it would be even more surprising if they were stationary. An overwhelming number of astronomers therefore accept the interpretation that the redshifts of distant galaxies are produced by the expansion of space.

If the galaxies are getting farther apart now, perhaps they were much closer together in the past. Even earlier in time, the entire observable universe might have been squeezed into a very small volume. Such a superdense state might have been the beginning of the Big Bang.

Hubble's ability to determine distances depended on his ability to calibrate the distance to standard stars. It is all well and good to know that one light bulb has twice the wattage of a second bulb, but if you do not know the actual wattage of at least one of the bulbs, you have a problem. Perhaps the bulbs are 100 and 200 watts, and they appear faint because they are far away from us. Perhaps the bulbs are 25 and 50 watts, and although they appear faint they are fairly close to us. The calibrations of standard stars are still difficult to make and, as a consequence, it is difficult to tell how much time might have elapsed since the Big Bang. The most astronomers are willing to say is that the universe appears to be somewhere between 10 and 20 billion years old.

The ages of meteorites can be determined by measuring the extent to which radioactive elements in the meteorites have decayed. The oldest meteorites, which astronomers believe formed with the Sun and the planets, are measured to be roughly 5 billion years old. (The oldest rocks

found on the Earth are 1 billion years younger than the oldest meteorites.)
The age of meteorites tells us that the inferred age of the universe is at
least in the right ballpark. (It would be disturbing if the calculated age
of the universe were less than the age of some of the rocks that make it
up.)

Because light takes time to travel, the notion of a beginning in time
requires the concept of the observable universe. The observable universe
is that portion of the universe from which light has had enough time to
reach us since the Big Bang. If 10 to 20 billion years have elapsed since
the Big Bang, we can see no farther than 10 to 20 billion light-years (this
is not quite accurate, but it is close enough for our purposes). The universe
might be infinitely large, or it might not be, but we can only see that part
inside a 10- to 20-billion-year "horizon."

Even a journeyman detective requires more than one piece of evidence before settling on an interpretation. Is there any other evidence,
corresponding to the windowsill without fingerprints, to corroborate the
story that the universe was created in a Big Bang 10 to 20 billion years
ago? Indeed there is. The second major piece of evidence supporting this
story was provided in 1965 by two American radio astronomers, Arno
Penzias and Robert Wilson. They were not seeking to corroborate the
Big Bang story—they were trying to study radio emissions from the Milky
Way using a very good radio antenna designed to communicate with one
of the first communication satellites. What they discovered was that they
could not eliminate all the background hiss that their antenna picked up
from the sky. This background hiss took the form of photons with energies
much lower than those of the photons Hubble observed. Furthermore,
these photons seemed not to be coming from individual galaxies but to
be uniformly distributed across the sky.

Unknown to Penzias and Wilson, this "cosmic hiss" was predicted in
1948 by Ralph Alpher, Robert Herman, and George Gamow, who con-
jectured that the universe was once in the form of a giant nuclear "egg."
The initial expansion of this egg was the Big Bang. As the egg expanded,
the chemical elements were created—first hydrogen, then helium, and

then the heavier elements. In the simplest model they investigated, however, all the hydrogen in the early universe would have been converted into helium. This would have been disastrous, because there would then be no hydrogen to fuel the stars.

By examining the energies of the photons emitted by luminous objects, astrophysicists can infer the nature of the matter emitting the photons. Each element has distinct fingerprints. When astrophysicists carry out analyses of a large number of objects in the universe, they come to a surprisingly simple result: most of the universe is made of hydrogen, and what isn't hydrogen is largely helium. (This finding seems a little odd to us, because hydrogen and helium are not very abundant on Earth, which is largely made up of carbon, oxygen, nitrogen, silicon, iron, and other "heavy" elements. The Earth is not typical of the universe as a whole, at least not as far as its composition is concerned.)

Alpher, Herman, and Gamow calculated that in order to account for the abundance of hydrogen observed in the present universe, a large number of very energetic photons would have had to be present to blast the helium nuclei apart as fast as they could form. As the universe expanded, these photons would have lost energy, and helium could eventually have formed, but not before the density was so low that not all the hydrogen could combine to form helium. Furthermore, the photons would have wound up in the present epoch with energies physicists associate with objects whose temperatures are only 5 degrees above absolute zero. There was a major problem with this model, however. Alpher, Herman, and Gamow could not figure out how this initial expansion produced elements heavier than lithium. As a result of this failure, most astronomers lost interest in the model. (Astronomers now believe that the heavier elements, such as carbon, oxygen, silicon, and iron, are forged in the interiors of stars.)

The energies of the photons discovered by the radioastronomers Penzias and Wilson were those associated with objects that are approximately 3 degrees above absolute zero—quite a remarkable agreement with the earlier, largely forgotten prediction. This radiation is remarkably uniform across the sky. So uniform that it provides convincing evidence for the uniformity of the universe at the earliest times, when the photons were very energetic.

We should not be too quick to accept this interpretation without asking whether some other interpretation for the cosmic background is possible. Some scientists have suggested that these photons might be emitted by very cool dust in intergalactic space. There are problems with this interpretation, however. If these photons did arise from emission from dust, the distribution in their energies would bear some imprint of the material that emitted them.

One of the most damaging blows to this interpretation was delivered by the COBE satellite—but not until 1988, more than 20 years after the cosmic background was discovered. COBE first measured the microwave background above the interfering effects of the Earth's atmosphere. The satellite revealed that not only are the cosmic photons remarkably uniform in their distribution across the sky (the only observed nonuniformity is exactly the sort that would be produced if the Milky Way were moving with respect to this background), but that the energies of these photons correspond exactly to the energies associated with photons that have not interacted with any matter since a big bang. Like the windowsill in our detective story, the energies of the cosmic photons display no fingerprints of any source such as cosmic dust.

The redshifts of the galaxies and the cosmic hiss provide the first two major pieces of evidence in support of the Big Bang story—the footprints in the mud and the windowsill without fingerprints. The third piece of evidence—the opened letter—comes from an examination of the nature of the matter that makes up the observable universe.

In the Big Bang model, the universe was originally pure energy. As the universe cooled, the lighter chemical elements formed from this energy according to Einstein's famous expression for the relationship between matter and energy. Exactly which elements formed and what their relative abundances should be follows from the present temperature of the cosmic background radiation and the average amount of matter in a representative volume of the present universe that includes both stars and intergalactic space. These calculations have been carried out in detail.

Calculations are one thing, observations another. By analyzing photons from a variety of astronomical objects, astrophysicists can estimate the present abundances of the lightest chemical elements in the universe. The range of abundances is vast, with roughly 100 million hydrogen atoms for every lithium atom, for example. (There is no intrinsic reason, however, why hydrogen and helium should be so much more abundant than deuterium and lithium.) Amazingly, the observed abundances of elements agree exceedingly well with the values calculated from the Big Bang model, representing the third major piece of evidence in favor of the Big Bang model—the opened letter.

The evidence for the Big Bang is the redshifts of galaxies, the cosmic-background radiation, and the observed abundances of the lighter elements. These three pieces of evidence come as close to representing a "beyond-a-reasonable-doubt" case for the Big Bang as anything you can find in astronomy—or anywhere else in science, for that matter.

It should now be clear why the Big Bang story is so firmly entrenched—like a three-legged stool, it is very difficult to upset. Any contending model must provide a plausible explanation not only for the redshift of the galaxies but for the cosmic background as well, *and* for the observed abundances of the lighter elements. No competing story has come close to filling this tall order.

Unlike in a detective story, in science the evidence is rarely complete. For example, there is no evidence that allows astronomers to infer with any confidence whether the expansion of the universe is slowing down, proceeding at a constant rate, or speeding up. Furthermore, the projected time since the Big Bang cannot be pinned down any more closely than to some time between 10 billion and 20 billion years. Finally, it seems unlikely we will ever know whether the universe is finite or infinite.

Nature is a truthful witness but one that rarely volunteers any information. Scientists must always interpret her answers carefully. Today's interpretations may not be exactly the same as tomorrow's. Nevertheless, the story of the Big Bang is one that most scientists would bet they will be telling for a long time to come.

This essay was based on the John L. Engelke Memorial Lecture given in February 1991 at the Twelfth Annual Darwin Festival at Salem State College, Salem, Massachusetts.

IS SCIENCE COMMON SENSE?

LAWRENCE E. JOSEPH

We had the sky, up there, all speckled with stars, and we used to lay on our backs and look up at them, and discuss about whether they was made, or only just happened—Jim he allowed they was made, but I allowed they happened: I judged it would have took too long to make so many. Jim said the moon could a laid them; well, that looked kind of reasonable, so I didn't say nothing against it, because I've seen a frog lay most as many, so of course it could be done. We used to watch the stars that fell, too, and see them streak down. Jim allowed they'd got spoiled and was hove out of the nest.

—*Mark Twain,*
Huckleberry Finn

Is science essentially commonsensical, like Jim and Huck making their best guesses about the stars? Or is its role more to expose the errors of simplistic, commonsensical assumptions, such as that the moon, which from a raft on the Mississippi looks plenty big enough, laid the stars like eggs? And when scientists reach conclusions that utterly defy common sense—such as that the moon, the stars, Jim, Huck, the Mississippi River and everything else on earth and in heaven all come from an ancient exploding ball no bigger than a mush-melon, as eminent Big Bang theorists have powerfully argued—how does science keep from getting heaved, like a spoiled star, out of the nest of ideas that levelheaded layfolk like to tend?

Common sense is the quality of judgment necessary to know the simplest truths, to recognize striking absurdities, and to be shocked by palpable contradictions. It is, as the sayings go, the ability to tell shit from Shinola, your butt from your elbow, chalk from cheese. It is at once a baseline requirement and a special attribute, a sine qua non that, in fact, many people lack. To say that someone has common sense is, therefore, to pay no small compliment; it implies a sharp eye for the significant, a grasp of the obvious—like seeing the emperor's no clothes—that at times can make everyone else seem color-blind.

The Oxford English Dictionary first defines common sense as "a sense common to all except lunatics and idiots," then gradually upgrades to "good sound practical sense: combined fact and readiness in dealing with the everyday affairs of life; general sagacity." Yet the more facts people know, or ought to know, the less commonsensical we may be apt to find those people. For example, Twain's heroes can be admired for their best guesses about the stars as long as we understand that they never had a chance to study astronomy. At the time the story is set, around 1850, the sun was still the center of the universe, at least in the cosmology popular among those with some formal liberal education. But Huck was probably too young and restless, Jim too deprived, and the local religion of the widow Douglas too God-starring to leave much room for Co-pernicus. In England, the famous Herschel family of astronomers were busy charting the Milky Way and beyond, though not, most likely, as far as Hannibal, Missouri. So the Mama Moon theory, natural and clever from Huck and Jim, would mark any informed scientist who adhered to it as a lunatic or an idiot.

In *The Common Sense of Science*, Jacob Bronowski suggests that sci-entists per se did not exist much before the time of *Huckleberry Finn*: "Science is not a special sense. It is as wide as the literal meaning of its name: knowledge. The notion of the specialised mind is by comparison as modern as the specialised man, 'the scientist,' a word which is only a hundred years old" (from the 1970 edition). Perhaps the divergence of science and common sense began with the increased use of technical instruments, a cause and effect of the Industrial Revolution that, like the Herschels' forty-foot reflecting telescope, extended scientific investigation beyond the sensory capacities of the common person. Or the rift might have been caused by the accelerating importance of mathematics, the very "lies, damned lies and statistics" that Twain could not abide.

Science and common sense may have parted ways even earlier, as Rafael López-Pintor of the University of Madrid argues: "The notion that science has an independent logic of its own is due to the vision of Newton and Descartes, i.e., that science's purpose is to discover the mental errors of 'common-sense' observations." The immortal French mathematician did have a flair for scientific snobbery, leaving the common folk to grind their Cartesian axes on his famous witticism "Common sense is the most equitably distributed thing in the world, since each man thinks he has it." And President Newton, looking down from high atop his Royal Society (perhaps through the first reflecting telescope, which he, of course, built), must have had a hard time believing that the particles lighting his calculating mind could possibly be attracted, gravitationally or otherwise, to the infinitely limited bodies below.

Scotsman Thomas Reid founded the school of common-sense philosophy with his *Inquiry into the Human Mind on the Principles of Common Sense* (1764), which was the first major philosophical text to challenge the Cartesian doctrine of universal doubt, i.e., that because human senses are fallible, our ideas, beginning with that doubt, are all that we can be sure of. Since common sense dictates that we trust and rely upon our senses in almost all circumstances, Reid reminded the skeptics that all reasoning, even their precious dubiety, originates with sensory impressions. His common-sense philosophy thus lent philosophical legitimacy to the scientific method of investigation, experimentation, and analysis.

Reid's countryman James Clerk Maxwell, one of the first great experimental physicists, realized science's debt to common-sense thinking. Teaching that today's science is tomorrow's common sense, Maxwell was an exponent of the experimental method and built the Cavendish Laboratory at the University of Cambridge on those pragmatic principles, starting in 1871. His theory that electromagnetic radiation is made of waves has, along with Newton's earlier assessment of light as particles, rippled through our common knowledge ever since.

Less charitably, Thomas Henry Huxley, the biologist and educator who was one of Charles Darwin's principal exponents, considered common sense the ape from which science has gloriously evolved. In 1868 he wrote, "Science is nothing but trained and organized common sense, differing from the latter only as a veteran may differ from a raw recruit: and its methods differ from those of common sense only as far as the guardsman's cut and thrust differ from the manner in which a savage wields his club."

From Social Darwinism to scientific socialism, science was steadily replacing common sense in the affections and affectations of midcentury intellectuals. "Science is a first-rate piece of furniture for a man's upper-chamber, if he has common-sense on the ground floor." Oliver Wendell Holmes warned in 1892, in *The Poet at the Breakfast-Table*. Perhaps Holmes felt a patriotic obligation toward common sense, well aware that his country's founding fathers had paid the ultimate homage to Scottish plain thinking through Jefferson's stubborn adherence to the philosophy and especially in Thomas Paine's *Common Sense*, the revolutionary polemic that argued the pointlessness of Britain's transoceanic rule over our emerging North American nation.

In France a ferocious school of Reid-inspired pragmatism led by C. Buffier, an eighteenth-century Jesuit turned atheist, had held such sway that common sense was installed in 1815 as the nation's official philosophy! But the Socialist upheavals of 1848 sent a shock wave of "scientific" and utopian political principles throughout Europe, and by 1870, beribboned old common sense was officially dethroned, left in the Republican dust.

Pity the poor sociologists, who have struggled since the founding of their discipline to prove that their work is really science, not just common sense. (Should sociology even be discussed in this collection?) Émile Durkheim, often regarded as the founder of sociology, believed that science is distinct from common sense and that sociology is a science. Durkheim's turn-of-the-century innovation was to apply the empirical and statistical methods of natural science to the study of society. A left-leaning proponent of the authority of the collective social mind and of the bonds of social order, Durkheim's legacy may be measured by the consensus of critics and supporters alike that sociology has less to do with common sense with each passing year.

(One wonders at the mixed emotions the great sociologist might have felt given the results of a 1986 survey taken by the Department of Sociology of Education at the University of Warsaw. Polling what it took to be a representative sample of the Polish citizenry aged fifteen to sixty-four, the social scientists examined "scientific and nonscientific knowledge in the social consciousness." The survey found that scientific knowledge was generally considered to be confined to a specific discipline, while nonscientific knowledge—artistic/literary, speculative, religious, and parascientific—was defined as common sense. The assumption that the more education people had, the likelier they would be to participate in science, was not substantiated. More than two-thirds of Polish society, the government survey found, sought the answer to the meaning of life in religion.)

Perhaps physics might have paused to blow a farewell kiss to common sense in 1905, Einstein's immortal year. Not much in special relativity, in the beginnings of quantum theory, and in the principles of subatomic motion resembled the working assumptions and rules of thumb by which common folk judged the physical world. However, certain notions, such as that time is the fourth dimension, really need never have been so mysterious.

Let's say that you are to be honored with a star-studded (but *très intime*) gala hosted by the Rockefeller family at their Center, and that you don't know your way around New York. What information do you need? The avenue (the GE building is really on Sixth, but enter from Fifth for the full promenade effect), cross streets (Forty-ninth and Fiftieth), and floor (just ask for the Rainbow Room). There you have your three dimensions, the x-, y-, and z-axes that high-school geometry taught are necessary to locate any point in space. But there's something else you need to know: When? What time? The *t*-axis, the fourth dimension. The day before yesterday?!

People kept their appointments before space-time continua were hypothesized and before the relativity theory, which explores the potential for movement along that *t*-axis, was proposed. (One might even suggest that for two and a half millennia physicists had been mesmerized by Euclid, by his timeless axioms and perfect planes. In Euclid's geometry, all truths are eternal—no need for niggling *t*-specifications. The first non-Euclidean geometry is usually considered the mid-nineteenth-century work of mathematicians Lobachevsky, Bolyai, and later Riemann, but in the journal *Philosophy of Science,* June 1972, Norman Daniels contends that Thomas Reid laid the foundations for a similar "commonsense" geometry some fifty years before his predecessors.) As Einstein's relativity roared into the twenties, common sense was, however unwittingly, already close enough for jazz.

As Newton's particles and Maxwell's waves coalesced into the quantum theory developed by Louis de Broglie and Niels Bohr, even Einstein became incredulous. The man who, in Bronowski's wonderful formulation, had asked himself what he would see if he rode around on an electron traveling at the speed of light, could not believe what he saw in quantum theory. The celebrated Einstein–Podolsky–Rosen electron-pair paradox, a thought experiment formulated over the course of the middle third of this century, challenges the doctrine of instantaneous nonlocal causation. Assume that two subatomic particles are created in a collision. Quantum physics holds that, in certain cases at least, the moment the spin of one member of that particle pair is determined by observation, the other member of that pair, no matter how far away it is, even halfway across the universe, will instantaneously—faster than the speed of light!—go from an indeterminate state to one in which its spin is known to be exactly equal and opposite to the spin of its mate. Heisenberg's "certainty principle," if you will.

Imagine two identical cue balls made at the same time and from the same material; one ends up in a Sandusky, Ohio, pool hall and the other is exported to Fiji. Were they to act like the quantum pair of subatomic particles, the moment the Ohio cue ball was shot and sent spinning clockwise the Fiji ball would be found to be spinning counterclockwise at exactly the same rate. This conclusion, affirmed in principle by Bohr, elaborated by the British theoretical physicist David Bohm, and recapitulated in J. S. Bell's theorem in 1966, stands as a tenet of quantum physics. At the subatomic level, common sense surrenders.

All around, it has been a very tough century for common sense, even among the philosophically inclined; most philosophers have dropped pragmatism like a dead cat. The exception is George Edward Moore, who gamely played loyal opposition to logical positivists led by Ludwig Wittgenstein, who some say sycophantically insisted on making his own work look as scientific as possible by numbering every statement—e.g., 1.23.33 or such like. Bertrand Russell did write under the banner of common sense, though more often than not his version of what's "CS" resembled his own atheistic, hedonistic humanism. Today, semioticians, deconstructionists, and allied critical theorists turn up their nose at common sense, which they consider a system for imposing majoritarian values on the minority, and therefore not politically correct.

With the promulgation of technology, a growing share of the educated public has had to acquire advanced quantitative skills. But as educator Lionel Elvin has pointed out, in this task common sense is rarely of much value: "The one real distinction is between mathematics and other studies. Mathematics relies for its validity on internal consistency alone; other studies involve in addition the test of correspondence with human experience" (*The Place of Commonsense in Educational Thought,* 1977).

When I attended Brown University, there was an infamous applied-mathematics professor absolutely dedicated to proving Elvin correct. The professor would stride into the lecture hall and say, "Given." He would then proceed to cover half the quadruplex chalkboard with equations, which students would scribble down dutifully. Eventually he would pause and say, "It is therefore intuitively obvious to the most casual observer that—," and scrawl equations over the rest of the board. There was nothing "intuitively obvious" whatsoever about that man's equation making. As for internal consistency, the students took that on faith. Sometimes at the very end he would add an exclamation mark, which should technically have been taken as a factorial indicator, but from him it was understood as a stylistic kind of thing.

Perhaps the most stinging denunciation of common sense has come from behavioral psychologists, who believe that human beings are flawed machines that can be improved by science: "The disastrous results of common sense in the management of behavior are evident in every walk of life, from international affairs to the care of a baby," wrote B. F. Skinner in *About Behaviorism* (1974). He went on to argue that "we shall continue to be inept in these fields, until a scientific analysis clarifies the advantages of a more effective technology. It will then be obvious that the results are due to more than common sense." His own efforts to create a "more effective technology" included the "Skinner baby box"—a large, air-conditioned, ostensibly germ-free enclosure with levers geared to provide rewards in response to appropriate actions—in which he attempted to raise his own daughter "scientifically."

Ideologues have their vanities. To the science chauvinists of the world, common sense means agreeing with the ideologues. But ironically, many of those who wholeheartedly accept the quantum implications for instantaneous nonlocal causality—because the theory of the forever-fated subatomic particles is Science with a capital *S*—tend to snicker at any suggestion, say, that there could ever be anything but a chance connection between a mother waking up frightened in her bed at about the same moment that somewhere out on the highway her child has died in a car crash. That, of course, is a folk theory, preposterous and naive.

Through its organ, *The Skeptical Inquirer,* the CSICOP (Committee for the Scientific Investigation of Claims of the Paranormal) doggedly showers cold water on notions of psychic phenomena, UFOs, telepathy, and anything else it deems "pseudoscience." The CSICOP describes its purpose as encouraging "critical investigation of paranormal and fringe science claims from a responsible, *scientific* point of view." Yet in "Science and Commonsense Skepticism" (Fall 1991), network software consultant

John Aach instructs his fellow enforcers that when scientific arguments fail to dissuade the public, common sense "seems especially well suited to deal with paranormal claims." In the CSICOP's case, and with science police generally, "responsible" means starting with the assumption of guilt, i.e., that those considered "fringe" or "paranormal" are wrong. Common sense is whatever induces the (benighted) lay public to reject unauthorized and therefore unscientific claims.

Artificial-intelligence researchers struggling to imbue their machines with common sense are finding their most stunning success in the field of humility instead. Marvin Minsky has long acknowledged that "common sense is not a simple thing. Instead, it is an immense society of hard-earned practical ideas—of multitudes of life-learned rules and exceptions, dispositions and tendencies, balances and checks." Minsky points out that, try as they might, scientists "are not doing very well at making common-sense machines that do the sorts of things ordinary people do. . . . [Computers cannot] understand ordinary language or look around a room and identify the objects visually. . . . There's still no machine that can look around and see where the chairs are and where the people are" (*The Society of Mind,* 1986).

Minsky's protégé Douglas Lenat heads up Cyc, a project with the express purpose of creating a computer capable of reasoning with common sense. Lenat envisions Cyc as being the "semantic glue" holding vast computer networks together, protecting man and machine alike from senseless computer goofs. Two person-centuries of work will have gone into Cyc before Lenat's team at MCC Corporation in Austin is through, at which point Cyc will be able to field a multitude of questions such as "On a rainy day, is it a good idea to carry an umbrella?" (The answer is yes.)

Perhaps computer can learn from insects. When I asked him if ants have common sense, Edward O. Wilson tossed off a definition good enough to cover a multitude of species: "If common sense means living by a set of rules of thumb that have worked well in the past, but living without examining those rules too closely or in detail, then, yes, ants have common sense. Collectively, but not individually." Wilson added that the careful strategic reactions of ant colonies in many ways resemble the responses that might be formulated by the human brain.

As for how scientists might keep from getting heaved out of common folks' nests upon presenting radically counterintuitive findings, one can find no more instructive example than James Lovelock. Lovelock, the British atmospheric scientist best known for his Gaia hypothesis arguing that the Earth is a homeostatic living organism, has labored for most of the past two decades to provide the biogeochemical and mathematical basis for what he believes to be the emerging science of geophysiology. Having also addressed the philosophical and spiritual ramifications that many discern in his Living Earth theory, Lovelock is sometimes asked his opinion on the nature of the universe: "I don't know. It's too big. There's quite a lot of it, actually. I'll stick to Earth," he responds.

Is science common sense? No, it's far beyond, and occasionally far askance. Does it need common sense? Not always, but sometimes desperately. Jim and Huck can't decide whether the stars were made or just happened. Even if scientists think they know, common sense says to make no final pronouncements about heaven and Earth until absolutely sure.

REDISCOVERING CREATION

LAWRENCE M. KRAUSS

Time past and time future
What might have been and what has been
Point to one end, which is always present.
—*T.S. Eliot,*
"Burnt Norton"

There is probably no more fundamental mystery in science, or in human imagination, than the nature of the origin of the universe. As self-awareness dawned in the first of our species, a growing puzzlement about Creation must not have been very far behind. As puzzlement gave way to wonder, and wonder to reason, human civilization remained obsessed with trying to understand how all we can see came about. The earliest organized writings, from the Sumerian *Enuma Elish* to the Sanskrit Vedas, from the Egyptian "Papyrus of Nes-Menu" to the Judeo-Christian Genesis, all focused on trying to make comprehensible, in very human terms, the mystery of Creation. And direct human experience suggested that if there was a Creation there must have been a Creator. And surely a Creator would not be content with merely the act of Creation, but would want a stake in the outcome as well . . .

Against this background, modern science emerged, more successful at problem solving, prediction, and control than anyone might have had any right to expect. Scientists have been just as obsessed with questions of origin and evolution as philosophers and prophets. At times, the vision of the world that emerged from scientific quest merged nicely with religious doctrine. At other times, it did not. When Descartes, rational philosopher and scientist, proposed a cosmogony designed to explain the observed motion of heavenly objects, a cosmogony containing the seeds of the modern law of physics known as *conservation of momentum,* he imagined that motion was *primordial,* given by God. On the other hand, when Galileo argued in favor of the Copernican notion that the Earth was not the center of the universe, the Church tried and convicted him for his views.

A marriage of convenience has existed between science and religion for the past four hundred years or so, since the time of Galileo. Scientific discoveries have been accepted by all but the most ardent fanatics. Recently, the Catholic Church officially pardoned Galileo and thereby officially accepted the heliocentric picture of our solar system. Confirmed believers have come to recognize that religion, which is based on *faith,* need not be incompatible with science, which is based on empirically proven or provable *fact.* Yet at the same time, as discoveries in physics and astronomy have pushed back the frontiers of the unknown, some have wondered just how far this process could go. Could the laws of physics explain all that we can see and all that we cannot, or would the "hand of God" eventually make itself manifest? Nowhere has this question been more potent than in the field of cosmology, the systematic study of the origin and evolution of our universe. Discoveries in this century have established, unequivocally, that the observable universe is expanding uniformly in all directions. It is most probable that the observable universe had its beginnings in an unimaginably dense, hot fireball, whose Big Bang explosion started a cosmic clock that has been ticking for almost 20 billion years, during which time the Earth, the Sun, the stars in our galaxy, and all the stars in all galaxies, formed.

Modern cosmology, as a true observational science, began only in the early part of this century, with the building of a new generation of large telescopes. It is easy to forget how far we have come in less than seventy years. Well after the other revolutions of twentieth-century physics had begun, most of the universe as we now understand it was not even known to exist. The question of whether there were other "island universes" beyond our own Milky Way was not settled until 1923, when Edwin Hubble, using the newly built 2.5 meter telescope on Mount Wilson in California, established clearly that the *spiral nebulae* (such as the one in Andromeda) are not part of our galaxy but are more than 750,000 light-years away. Our sun is about 8 light-minutes away (if the Sun exploded as you were reading this, you wouldn't know about it for 8 minutes) from the Earth. The nearest star other than the Sun is about 3 light-years away. Our galaxy is about 100,000 light-years across, and the distance to the nearest large galaxy outside our own is about 1 million light-years. We now estimate that there are at least 100 billion galaxies, more or less like our own, in the visible universe, containing on average 10 billion to 100 billion stars each! While galaxies themselves are often clumped together into clusters and superclusters (our own galaxy is on the outskirts of the Virgo supercluster), there seem to be roughly the same total number of galaxies in all directions as we look outward.

Hubble made another, more important discovery with the new Mount Wilson telescope. Over the course of ten years he estimated the distances to over two dozen galaxies and discovered that on average their velocities away from us are directly proportional to their distance from us: galaxies twice as far away as others are receding twice as fast. In the course of the last sixty years Hubble's observations have been refined, confirmed, and shown to be universal. The visible universe is expanding uniformly in all directions.

The idea that the universe is expanding has dramatic consequences for our picture of both Creation and for the eventual fate of the universe. The Hubble expansion implies that the universe as we know it has not been around forever. *The visible universe had a definite beginning!* If the observed expansion is extrapolated backward—about 10 to 20 billion years—all of the objects we see today are contained in an incredibly dense, single region smaller than a pinhead.

It is hard for anyone, cosmologists included, to picture such a state, which is perhaps why this theory was not taken very seriously until about 1965 when a remarkable accidental discovery gave the Big Bang theory a firm foundation. If the visible universe began in a hot Big Bang, the laws of physics predict that as it expanded and cooled, there would have been important milestones at certain temperatures. When the universe was one second old, at a temperature of around 10 billion degrees, nuclear reactions would have occurred in profusion, causing the eventual formation of the nuclei of almost all of the light elements, including hydrogen and helium, that make up stars today. (The remainder were produced in the stars themselves.) Much later, when the universe was about 300,000 years old, its temperature would have cooled to about 10,000 degrees, so that neutral atoms could form as positively charged atomic nuclei captured negatively charged electrons to orbit them.

Once neutral atoms formed, matter would essentially have stopped interacting with the radiation present at that time. As the universe continued to expand this radiation would have gone on cooling, moving unobstructed through the universe. Today, after almost 20 billion years, this ubiquitous radiation would have cooled to about 450 degrees below zero degrees Fahrenheit, or about 3 degrees above the lowest temperature anything can have, absolute zero. A background of electromagnetic radiation at this temperature would exist in the form of microwaves similar to the waves used for microwave communication on Earth.

Enter Arno Penzias and Robert Wilson. In 1965, these two young scientists at Bell Laboratories in New Jersey decided to use sensitive microwave antenna on the grounds of the lab to try to do radio astronomy. They discovered instead an irremovable background noise in the antenna. Even after they removed a "white dielectric material," in their words— or pigeon droppings, as they are more familiarly known—the background noise persisted. Frustrated, they went down the road to Princeton University to talk to some colleagues there. They learned that this "noise" they couldn't get rid of was probably the most important radio "signal" ever to be received from the heavens. It was the microwave "afterglow" of the Big Bang.

The *Cosmic Microwave Background* (CMB) provides a direct "picture" of how radiation was distributed throughout the universe when it was less than 1 million years old. This picture appears because this radiation has been moving through the universe largely without interacting with the intervening matter. The signal now being received originated at least 10 billion years ago. Moreover, because, at the time that the observed CMB was created, the radiation was interacting strongly with matter, the way the signal is distributed across the sky now can reflect the distribution of matter in the universe at that time.

Both the uniformity and the thermal nature of the CMB attest to its primordial origin. If this background were local to our galaxy, we would expect a signal that peaked in the direction of the galactic center or that at the very least varied in ways characteristic of the distribution of objects in the Galaxy. Instead, it is uniform in all directions, and can be described throughout by a single temperature. Just as the light emitted by an oven heating element tells us how hot that element is, the CMB is also thermal, that is, associated with a particular temperature. Though other physical processes might create background radiation, only a uniform Big Bang explosion naturally predicts that it will have a single universal temperature today—about 3 degrees above absolute zero if the CMB has been cooling ever since the Big Bang over 10 billion years ago. Both of these features of the CMB were confirmed beautifully by the COBE (COsmic Background Explorer) satellite launched by NASA in 1989. After twenty-five years of astronomers' tentative ground-based observations, COBE confirmed, *in the first nine minutes of its flight,* the CMB temperature to an accuracy of 1 part in 1000. Measurements in the first year of operation then showed that this background was uniform, not varying more than 1 part in 10,000 across the entire sky.

If the CMB provides us with a mirror of our cosmic past, telling us that the Big Bang expansion provides an accurate description of what the universe has been doing at least since it was 300,000 years old, it may also provoke more questions than it answers. Why was the early universe so uniform? How far back does the Big Bang picture remain valid? What established the initial conditions? What happened *before* the Big Bang? Will the observed expansion go on forever? Over the past 25 years, cosmologists have tried to address these issues using all of the theoretical tools at their disposal. Exotic phenomena in the very early universe, soundly based on the standard model of elementary particles developed over the last 20 years, have been invoked to explain everything from the observed uniformity of the CMB to the fact that most of the mass of the universe appears to be invisible to telescopes. If these theories are correct, then more than 99 percent of the mass of the universe is made of some new type of matter unlike anything we see on Earth. The combined gravitational attraction of all of this material will slow the expansion of the universe gradually, but never completely, so that the radiation background will continue to cool, the stars will eventually all burn out, and a cold, dark universe will get colder and darker for all eternity. . . .

This is where our theories of the cosmos stood on April 21, 1992, when George Smoot, head of one of the COBE instrument teams, made a historic announcement that reverberated in headlines throughout the world: The COBE satellite had discovered "ripples" in the microwave background. The COBE "April Surprise" sent shock waves throughout the scientific community as well as the popular press. Why should such an odd and seemingly simple observation cause such a fuss? Perhaps because Smoot accompanied the announcement with the remark that it was like "staring at the face of God." Indeed the ripples—small temperature fluctuations at a level of only slightly more than 1 part in 1 million—observed by the COBE satellite probably take us as close as we will ever come to witnessing the creation of our universe.

Special relativity tells us that it is impossible to move matter around at speeds greater than the speed of light. Yet the small deviations from uniformity observed in the microwave background exist on scales so large that no process from the beginning of the observed Big Bang expansion to the moment the CMB was created would have had time to build up such a signature. The largest-scale features observed would not have had time to develop even in the time since the CMB creation up to today. The simplest, and perhaps the only, consistent explanation of such a signal is that these structures were imprinted in the Big Bang expansion at the very beginning.

Is this then the smoking gun, the "hand of God" that many have awaited? Must we take these initial structures as given, beyond the laws of physics, and use physics to work only forward? Perhaps the most surprising thing of all about the COBE discovery is that the answer to this question is probably no. For it turns out that the very ripples that were observed by COBE arise naturally and unambiguously in what has become the standard picture of the early universe, based on ideas from elementary particle physics. A theory called *inflation,* originally proposed by Alan Guth in 1980, suggests that at the very earliest moments of the Big Bang, the universe could have undergone a sudden, cosmic burst of growth—increasing the size of all regions by a factor of more than a billion, billion, billion, billion times in less than a fraction of a second. This theory not only explains how the observed universe, as it settled down after the inflationary phase, could have become so isotropic but it also predicts that on top of this isotropy small fluctuations of a characteristic type would have been imprinted. It is exactly such fluctuations that the COBE team suggests its satellite has detected.

Moreover, the general features of the ripples observed in the CMB are consistent with the possibility that the gravitational pull of the density fluctuations associated with these ripples caused matter slowly to begin to collapse on different scales, forming all of the structures we observe in the universe today, including stars, galaxies, and clusters of galaxies—*everything* we see in the universe. But this is likely to be the case *only if* the universe contains 99 percent dark matter, which is what we would expect if a sudden inflation of the universe actually took place during the earliest moments of creation.

Cosmologists are now left with an embarrassment of riches. The simplest model we have invented to describe the universe, a model involving exotic phenomena that stretched the limits of credibility for some, predicts more or less exactly what has just been discovered. Of course much work remains to be done to check and complete the picture. The challenge now is to "find" the dark matter and to use terrestrial machines to uncover evidence of the elementary particle physics required to make something like inflation happen in the early universe. No doubt, more surprises await. . . .

But what does all of this imply for our view of Creation? It is possible that we will never be able to empirically address the issue of what happened before the Big Bang and why it began at all. If the inflationary theory is correct, any remnants of such an early time would have been erased during inflation. There is perhaps still room for the hand of God, at $t = 0$. But what we have learned from COBE is that the laws of physics are probably adequate to explain this grandest of cosmic mysteries as far back as the evidence—the earliest evidence we may ever be privileged to obtain—can carry us. While observations from this tiny satellite orbiting the Earth are in no way in contradiction with divine Creation, they push the epoch in which a supernatural hand of God may have bypassed the laws of nature (that the Creator herself may have established) back outside the realm of the observable. Belief in a Creator must remain a matter of faith, not scientific deduction.

In the Beginning

HAROLD MOROWITZ

Life on Earth, which had its origins well over 3.5 billion years ago, has been fruitful and has multiplied, filling every niche with single-celled bacteria and the more complex higher forms, including all manner of plants and animals and protists and fungi. This branching of the tree of life has led to an abundant flowering of species, wondrous experiments ranging from tiny viruses to great blue whales and giant redwood trees.

A few hundred thousand years ago, nature performed the most astounding trick of all: life gave rise to mind. With this advent of thought, the universe was suddenly able to turn inward and think about itself. Biochemist and natural philosopher George Wald once noted that a physicist is the atoms' way of thinking about atoms.

Slowly at first, and then at an ever-increasing pace, mind gave rise to civilization, and civilization gave rise to learning. For the past six hundred years, with the spread of the noosphere—the world of mind that has emerged with the evolution of *homo sapiens*—the global mind has taken over the planet. Humankind now possesses a wide range of tools with which to ask the eternal questions of who we are, where we have come from, and where we are going. I celebrate our ability to frame such questions, and I choose to participate in the party by asking how life on Earth began.

Since I hear no voice out of the whirlwind telling how the foundations of Earth were laid, I must use that most wonderous gift of the cosmos, thought, to probe life's origins. Through a study of today's myriad organic forms, the powerful generalizations of molecular biology and biochemistry, analyses of fossil remains, and our always changing understanding of evolution, we can try to reconstruct the properties of the universal ancestor, those earliest cells from which all life is descended. We can also start with our understanding of the geophysics and geochemistry of the early planet and postulate how the universal ancestor could have come to be. The extent to which the results of these two investigations are consistent is the measure of our success in understanding life's origin.

In the laboratory, our attempt to reconstruct the chemical process of biogenesis piece by piece permits us to prune away false hypotheses and reinforces our confidence in other conjectures. No view stands with certainty. Scientific research is an ongoing act of commitment, an attempt to understand the workings of the cosmic intelligence that is manifest in us. The laboratory bench, the computer terminal, and the desk are altars where we praise the design of the universe.

Thus, when I set out to retell the story "In the Beginning," it emerges not as a scientific journal article nor as a folktale, but as a scenario, a script for the cosmic drama. It is, however, a scenario crafted according to a number of firm rules:

Rule 1. Thou shalt not violate the laws of physics and chemistry, for these are expressions of divine immanence.

Rule 2. Thou shalt not trespass the Razor of Occam and multiply hypotheses, but shalt formulate the simplest of stories.

Rule 3. Thou shalt adopt a principle of continuity so that each stage of the grand scenario connects with the preceding and succeeding stages.

Rule 4. Thou shalt eschew miracles, for as Spinoza taught, they contravene the lawfulness of the universe.

In the beginning, the Solar System in which we live condensed out of the debris of stellar explosions mixed with primordial hydrogen from an earlier beginning, the Big Bang. Although the swirling system seemed unformed and void, the conservation of angular momentum, the force of gravity, and the forces of electromagnetism led to the condensation of a large central mass and a number of smaller masses orbiting the parent body. The pull of gravity accelerated all the material of the system toward regions of accumulating mass, releasing vast amounts of potential energy as heat. The central Sun, being the largest of the bodies, gave off the most heat, raising its temperature until it began to glow. And thus there was light to illuminate the entire Solar System. Thermonuclear fusion reactions then began in the glowing Sun to assure the continuance of the light for many eons to come.

As the Sun began to glow, the Earth, which was heated by its own gravity and the energy of its own radioactive decay, began to undergo a meltdown. The denser molten iron and nickel moved to the core, and the less dense silicon-, aluminum-, and oxygen-containing minerals moved toward the crust.

For the first 500 million years, the Earth continued to accrete material, not continuously, but in the episodic capture of meteorites and comets. These celestial objects ranged in size from tiny particles up to massive bodies 500 kilometers in diameter or larger. The biggest of the meteors possessed enormous amounts of potential and kinetic energy, which on impact were released as heat.

During the later part of the accretion epoch, the surface of the planet periodically cooled, and gaseous molecules were emitted from the crust and became oceans and atmosphere. The episodic giant-meteoritic impacts produced huge clouds of steam surrounding the planet. After the last great meteor collided with the Earth about 4 billion years ago, the steam condensed and rained down to form a permanent ocean. And the atmosphere was a firmament to separate the waters of the oceans from the waters of the clouds.

In the waters of the surface and in the atmosphere were a number of carbon compounds. Chief among these were the carbon dioxide of the atmosphere and the carbonates of the ocean. Comets, celestial dust, and certain meteors named carbonaceous chondrites also brought carbon compounds to the surface of the planet. Among these aggregates of carbon carried from the heavens and bubbling up from the interior were molecules called alkanes, oily substances made up of chain and loop molecules containing structures consisting of units of two atoms of hydrogen and one atom of carbon each.

One of the basic rules of chemistry known to every schoolchild and every chef is that oil and water do not mix. This rule also has exceptions, such as: soap removes grease from soiled clothes by floating it away in water, and egg yolk combines the oil and water of mayonnaise into a smooth mixture. These exceptions always involve structures called *amphiphiles,* which have the property of partitioning so that one end of the molecule is in oil and one end in water. *Amphiphile* means "loving both"—oil and water. Among the molecules brought to the surface of the early Earth, or made there by the action of the Sun's ultraviolet light on molecules of terrestrial origins, were these amphiphiles.

Amphiphiles are strange entities; they are not completely at home in either oil or water. In the waters of the ocean, when these molecules collide the oil-seeking parts adhere to each other, while the water-seeking parts interact with the aqueous surroundings. This results in collections of amphiphiles called *coacervates* (from the Latin meaning "to heap together"). Membranes are coacervates made of amphiphiles bonded together in sheets two molecules thick, the oil-seeking ends forming the interior of the structure and the water-seeking ends the exterior. We name these membranes *bimolecular leaflets* because they measure two molecules across. These spontaneously forming entities are a core structure of life—like all such structures, a gift of the laws of nature.

In obedience to the inexorable laws of physics and chemistry, biomolecular leaflets have one more remarkable property. In water they spontaneously form into closed shells called *vesicles.* Thus when the first bimolecular leaflet formed and arranged itself into a vesicle, something radically new entered the world. The membrane became a barrier to separate the interior waters of the vesicle from exterior waters. The partitioning of the inside from the outside is the beginning of individuality. In some vague and primitive way, the distinction between the I and the Thou was beginning to appear.

And so the waters swarmed with a myriad of vesicles, tiny protocells on their way to life. Among the vesicles were those that had dissolved chromophores within their membranes. These structures had existed among the primordial molecules. Chromophores are molecules that can capture light from the Sun and convert it into some other form of energy, such as the electrical potential between the inside and the outside of the vesicle. The vesicles with these kinds of chromophores became photo-electric cells. Among these primordial globules were some that, because of the presence of other molecules within them, could use the energy of light to drive chemical reactions. The continuous flow of energy is essential to all life. Photosynthesis thus entered the world. For those vesicles that had this energy-conversion property, there was a vast difference between the light of day and the dark of night. And there was evening and there was morning of that day.

As the result of chance inclusion of different molecules when the closed shells formed, the photosynthetic vesicles differed from each other in the chemical compounds they could synthesize. Certain rare vesicles were able to synthesize both amphiphiles and chromophores using the light of the Sun to drive the reactions. These grew more membrane material, and became larger. Large vesicles became unstable and spontaneously divided into smaller vesicles. Thus certain kinds of these primordial cells began to predominate by using the energy of the Sun to carry out the synthesis of more vesicle material of a similar kind. Replication entered the world.

While all of these light-driven chemical reactions were taking place, there was synthesized an occasional rare molecule that could stick to the membrane surface and catalyze one or more of the other reactions taking place in the vesicle. Thus catalysis entered the world. It was an emergent property, one that followed from the laws of physics and chemistry but required a certain level of organization before it became manifest. Those molecules that catalytically stimulated more rapid chemical reactions led to faster growth of the favored vesicles in which they were found.

A catalyst is a molecule that speeds up the rate of a reaction but emerges unchanged from that reaction. Modern-day catalysts in biological systems are usually proteins and are called *enzymes,* although ribozymes, catalysts made of the genetic material RNA, have recently been discovered. They are highly specific as to which reactions they accelerate, and they are very efficient. Primordial catalysts were undoubtedly much cruder and less efficient than their present-day counterparts, but they did speed up certain reactions and so were essential to the development of living cells from metabolic globules.

The earliest vesicles were thus characterized by a network of chemical reactions. This array of related processes also had emergent properties. The whole network was truly greater than the sum of its parts. It was the metabolic basis of life.

On occasion, a newly formed catalytic molecule was able to alter the network so as to lead to an increase in the amount of a chemical synthesized. This phenomenon, called *reflexive autocatalysis,* introduced molecular memory into the world. Memory is a necessity if some replicating systems are to be different from other replicating systems. In today's world, molecular memory is the province of an elaborate genetic system centering on very large molecules of DNA. The early replicating vesicle remembered rare mutant molecules by synthesizing more such molecules. The offspring of this vesicle, produced by synthesis and division, were different from the offspring of other vesicles, and this difference persisted. Thus speciation entered the world, for each such vesicle produced offspring after its own kind.

The most favored vesicles grew faster than the others, and inequality and competition inevitably entered the world. For surely the resources of a finite planet are finite, and each species of vesicle could not continue to grow exponentially in such a world. Good and evil entered the world of vesicles. For the good of the growth and evolution of one species of vesicles came at the price of the privation and perhaps extinction of another species of vesicle. Surely this, too, is an inexorable law of cosmic intelligence.

Just as large vesicles divided, so on rare occasions small vesicles fused. When the partners of this union were from different species of vesicle, new species formed, for they possessed the molecular memories of both parents. On occasion this gave rise to chemical networks superior to those found in either parent, and the new species began to outgrow all other species in its environment. This combination of remembered mutant molecules was such an effective mechanism it later became the biological norm in sex. For the primordial world, simple fusion sufficed.

Favorable properties leading to greater fitness began to accrue in more-and-more-elaborate networks, each network with its own metabolism and each species of vesicle in competition with other species. And so metabolism became central to life. For from metabolism come the appropriate forms of energy needed to build up orderly molecular structures and counter the degradation due to the second law of thermodynamics.

Thus far, functional memory was a property of networks. Some networks began to develop an alternative mechanism, memory molecules that could serve as templates for other molecules that played a catalytic role in the cell's metabolism.

Long before there was life as we know it, a great evolutionary branching of species of vesicles with memory molecules took place. More and more elaborate and sophisticated networks and memory-storage devices developed. One pathway to memory and metabolism eventually proved to be far more efficient than the rest. Along this evolutionary pathway there developed a species of vesicle that outcompeted all others. This species was the universal ancestor, for its descendants also speciated, but they preserved the intermediary metabolism and method of storing information that emerged from primordial evolution. All life on earth—grasses and fruit trees and fowl and beasts of the field—carries within every cell the metabolism and molecular genetics of this universal ancestor.

It has come to pass that for almost 4 billion years the essential chemical processes of life have remained the same. Mountains have risen and eroded away; continents have migrated over the surface of the planet; magnetic poles have shifted. All manner of geological changes have occurred. Through it all, the basic chemistry of life has remained fixed since the original ancestor. It is the most permanent feature of the Earth that we know. Therefore, the biogeochemical beginning is still within us: our cells are vesicles wherein the interiors are separated from the exteriors by amphiphilic bilayer membranes. These membranes are a tangible reminder of the beginning and the interrelatedness of all things.

And so it was that primordial chemistry begat amphiphiles, and amphiphiles begat vesicles, and vesicles begat pyrophosphates, and pyrophosphates begat keto acids, and keto acids begat amino acids, and amino acids begat nucleic acids, and nucleic acids begat the genetic code. And chemicals formed the first cells, the universal ancestor. This is the first book of generations.

OF TRACKS AND THE RIVER

JAMES O. FARLOW

The rising sun first touches the heights of limestone cliffs along the north side of the river, where the river makes its hairpin turn. As the sun gets higher, it starts to clear the tops of trees on the river's east bank, painting dead cedars that line the lower bluffs of the west bank a bloody red.

The hordes of tourists and school groups that will later crowd the river valley are still hours away, but I've no lack of company. I'm surrounded by calling cardinals and chickadees, and the chattering of a kingfisher echoes along the valley walls. A large fish leaps from one of the deeper pools, its splash momentarily startling me. The weather is colder than I thought it would be, and I shiver a bit and take another sip of hot coffee.

Finally the sun is high enough to fall on a clear, quiet patch of water right in front of me. Deep, oddly shaped depressions are visible in the carbonate rock of the riverbed. Some of them are like gigantic bird tracks, three-toed impressions that might have been made by the thunderbirds of Native American mythology. Other, even larger tracks look like the spoor of some monstrous cross between an elephant and a tortoise.

The river in question is the Paluxy, a stream located about fifty miles southwest of the Dallas–Fort Worth metroplex. The Paluxy River flows eastward across the Texas countryside to its confluence with the larger Brazos River east of the town of Glen Rose. Just west of Glen Rose, however, the little river bends to the north, then turns sharply back to the south. It is in this northern loop of the Paluxy River that one of the most spectacular concentrations of fossilized dinosaur footprints in the world is exposed.

The first tracks to be reported from this region were found in 1908, in a small tributary of the Paluxy, by a truant schoolboy named George Adams. These tracks were of the three-toed variety. Although some local people figured them to be footprints of giant birds, young Adams's schoolteacher recognized them for what they were, the tracks of bipedal, or "two-legged," dinosaurs. Similar dinosaur tracks were soon found in the bed of the Paluxy itself, and these tracksites came to the attention of Ellis W. Shuler, a geologist at Southern Methodist University. Shuler wrote a series of short technical articles about the Glen Rose footprints he had seen, most of which had probably been made by carnivorous dinosaurs. Soon after Shuler's first paper was published, dinosaur tracks began to turn up at numerous other sites across central Texas.

In the summer of 1934, footprints of a new type were found in the rock of the Paluxy's bed. These new prints had clearly been made by a four-footed, or quadrupedal, animal even larger than the creatures responsible for the three-toed tracks. The finders of the new tracks, two brothers named Moss, made the perfectly reasonable guess that these were the traces of some prehistoric elephant.

Four years later, however, their true identity was determined by a professional dinosaur hunter. In 1938, Roland T. Bird, a fossil collector for the American Museum of Natural History in New York City, found himself at an Indian trading post in Gallup, New Mexico, at the end of a summer spent searching for dinosaur bones in the American West. In the store's window were displayed two blocks of stone in which could be seen what looked like the footprints of a human giant. Upon inquiry Bird learned that the same trader had three-toed dinosaur tracks at another trading post, and that both the human and dinosaur tracks had come from Glen Rose, Texas. To Bird's eye there was an "artificial" look to all of these footprints, but he thought he should check out the place where they'd come from nevertheless.

Bird's suspicions were well founded. A few residents of Glen Rose had been supplementing their income by cutting dinosaur footprints from the bed of the Paluxy River for sale to tourists, and some folks had simplified the process by not even bothering to find real tracks first, but instead simply carving footprints into rock. The tracks Bird had seen in New Mexico had in fact been manufactured by none other than the now-grown-up George Adams. The footprint carvers insisted, however—as did other Glen Rose residents—that the faked human footprints, like the bogus dinosaur tracks, were based on real originals that could still be seen in the bed of the Paluxy. Bird asked to see some of the real "man tracks," but was unimpressed and disappointed with what little his guides were able to show him.

In contrast, Bird was very impressed by the three-toed dinosaur tracks in the Paluxy's bed, and even more excited when he learned about the larger footprints seen by the Moss brothers. Bird's boss at the museum had in mind the possibility of constructing an exhibit of dinosaur footprints from various parts of the U.S., and Bird thought that the Paluxy might have quite a bit to offer such a display.

He poked around the riverbed in several places and was rewarded for his efforts by finding some examples of the huge, elephantlike tracks. Bird recognized, however, that these were the footprints not of elephants but rather of sauropods—a suborder of plant-eating dinosaurs (informally called brontosaurs after *Brontosaurus* itself, an animal more correctly known as *Apatosaurus*) that would have dwarfed the biggest elephant. This was so exciting a discovery that in 1940 Bird returned to the Paluxy River to collect sections of a sauropod trail, as well as part of the nearby trackway of a carnivorous dinosaur that had seemingly been following the big herbivore, thoughts of dinner perhaps on its reptilian mind. One segment of the two trails was eventually reassembled at the American Museum of Natural History, beneath the mounted skeleton of a sauropod, in such a way as to give museum visitors the impression that the skeleton was a living animal caught in the very act of making tracks. Another portion was put on display at the Texas Memorial Museum in Austin.

I didn't become concerned with dinosaur footprints until forty years after Bird did his work in the Paluxy. In the winter of 1980, when I was

—

teaching in the geology department of a small college in Michigan, I received a letter from a college friend who'd remembered that I had an interest in dinosaurs. He had married into a family that owned a ranch in Texas about 150 miles southwest of Glen Rose. A flash flood had recently stripped tons of stream gravel and shingle from a draw on the ranch, revealing a large assemblage of three-toed-dinosaur footprints in the limestone floor of the usually dry creek. My friend wanted to know if I was interested in taking a look at the site.

As it happened, another geology professor and I were scheduled to lead a field trip of students that coming spring, and we decided to take our group to the ranch. We spent a week mapping, measuring, photographing, and casting the dinosaur tracks and in general having a splendid time.

I had no intention of starting a major study of dinosaur footprints, and I didn't know very much about the scientific literature dealing with these fossils. I was aware, of course, of Bird's much earlier work along the Paluxy River; I knew that he had written a few popular accounts of his footprint-collecting activities. I was sure, however, that by now someone had to have written a long, detailed description of the Texas dinosaur footprints. All I had to do was find it, use it to identify the tracks from the ranch, and then write a short description of the site for publication in a paleontological journal.

It didn't work out that way. I soon learned that, apart from some short pieces written by Wann Langston of the University of Texas, no paleontologist had taken much professional interest in the dinosaur footprints since Bird's time. If I wanted a detailed identification of my tracks, I'd have to make it myself. I had other research projects that were of more immediate interest, and so, apart from publishing a short note on the significance of some of the ranch's dinosaur trails for arguments about the running ability of dinosaurs, I shelved the project.

A year later I changed jobs, joining the faculty of the geology department of the Fort Wayne campus of Indiana University. I knew that I'd be coming up for tenure in a few short years, and I knew that I had to publish some professional articles during my probationary period if there was to be any hope of getting tenure. Which of my various research projects was most likely to yield publishable results in the amount of time I had? After a bit of admittedly somewhat cynical thought, I decided it was time to dust off the footprint study.

By this time I knew that there were dinosaur footprints at sites across much of central Texas, but I also knew that for the number and quality of preservation of dinosaur tracks, the Paluxy River would be hard to beat. Even though I would ultimately visit tracksites all over the state, I would return again and again to Glen Rose and spend a lot of time in the river.

A day in the field begins early. I usually get up before sunrise and try to get to the river by first light. There aren't many places to eat in Glen Rose at five-thirty in the morning (or at any other time, for that matter), but Linda's is open by then. You can sit in a booth and eavesdrop on the talk of farmers and ranchers, or of workers at the nearby nuclear-power plant.

I remember once hearing a conversation between a waitress and two customers, the latter obviously out-of-towners; the waitress told them that they should be sure to "go to the park and see where the dinosaurs crossed the river." The park is Dinosaur Valley State Park, established by the state of Texas to protect the northern loop of the Paluxy River, in which so many dinosaur tracksites, including R. T. Bird's footprint quarry, are located. The waitress was stating a common misconception— but a reasonable one, from a geologically unsophisticated point of view. Since dinosaur footprints occur in the riverbed, and not on the dry land on either of its banks, the idea that the tracks were made by fording animals seems hard to fault.

Even so, the dinosaurs didn't cross the river; they preceded it. Dinosaurs lived during the Mesozoic Era, which began some 245 million years ago and ended about 65 million years ago. Geologists divide the Mesozoic into three periods. The first of these is the Triassic period, after which comes the Jurassic period, and then the Cretaceous period. The dinosaurian track makers of Texas lived during the middle of the Cretaceous, just over 110 million years ago. This means that the trackmakers lived roughly 40 million years after such well-known dinosaurs as *Allosaurus, Diplodocus, Apatosauru,* and *Stegosaurus*—dinosaurs whose bones are common at Dinosaur National Monument, along the Utah–Colorado border—and about 40 million years before such equally familiar dinosaurs as *Tyrannosaurus, Triceratops,* and *Ankylosaurus.*

For a long time before dinosaurs left their tracks in coastal muds of the region that would one day be central Texas, this area was dry land. During millions of years of exposure above sea level, the land surface had been eroded to a fairly flat plain. Over the course of the early part of the Cretaceous period, the sea began an intermittent incursion across the land. Sometimes the sea advanced and then retreated, only to advance again and flood even more of the landscape. Eventually the expanding waters of this ancestral Gulf of Mexico would join a similarly advancing seaway that had come from the Arctic region, splitting the North American continent into eastern and western portions.

Generations of geologists have studied the rocks that formed in central Texas from sediments deposited by the sea and by rivers flowing into the sea. In the late 1800s, the basis for our understanding of the geology of this region was established by Robert T. Hill, a rather acerbic fellow who worked at various times for the U.S. Geological Survey, the University of Texas, and the Texas Geological Survey. Subsequent workers have extended and refined Hill's concepts.

The slowly advancing waters of the Early Cretaceous Gulf of Mexico moved forward and backward across the Texas landscape in three major episodes. Sedimentary rocks that formed during the first advance and retreat are called the Trinity group. The Trinity group itself records three smaller-scale advances and retreats of the shoreline; rocks formed during the last of these constitute the Twin Mountains formation (also known as the Hensel sand and the Bluff Dale sand) and the Glen Rose formation. Research conducted by Jeff Pittman of the University of Texas indicates that dinosaur footprints occur at numerous sites, concentrated at two major levels, in the Glen Rose formation. The footprint sites of the Paluxy River occur at the lower of these track levels.

At the time the footprints of the Paluxy River were being made, the

Gulf extended much farther inland than it does now. The overall environmental setting of the Glen Rose region was something like that of the modern Georgia Sea Islands, but also much like those of the Everglades and the Florida Keys. A line of reefs, constructed by corals and a peculiar group of extinct clams, was located offshore. In shallow marine waters on the landward side of the reefs lived a diversity of sharklike and bony fishes, along with a host of invertebrates, including worms, clams, snails, and sea urchins. As in the Florida Keys and the Bahamas today, great quantities of lime were being made by seaweeds and microscopic algae; when these organisms died, lime particles were released as sediment grains that would eventually become carbonate rocks—limestones and dolomites. At the shoreward edge of the sea were extensive tidal mud flats, and also groves of ancient conifers that seem to have been able to live partly submerged in salt water, just as salt-marsh grasses and mangroves do today. Fossil pollen indicates that early flowering plants grew in the coastal region as well.

Farther inland, to the northwest, were forests of seed ferns and conifers, as well as more open habitats covered by true ferns. Bones and teeth of fishes, frogs, salamanders, turtles, lizards, crocodiles, large and small plant-eating and meat-eating dinosaurs, and small primitive mammals have been found in sandstones and shales that originated as sediments deposited by rivers that flowed across these terrestrial environments toward the sea.

Dinosaur footprints commonly occur in sedimentary rocks that formed within and just above the tidal zone, at the boundary between terrestrial and marine environments. Wide expanses of seashore occasionally became exposed for extended periods of time and then flooded, as seawater levels fell and rose in cycles greater in magnitude than the usual tidal rhythms. During periods of exposure, dinosaurs walked across the mud flats, sometimes leaving footprints in such abundance that the sediments were thoroughly churned by the trackmakers' feet. At some point after they formed, the tracks were filled by fine-grained sediments that were probably carried into the coastal areas by rivers, perhaps after heavy rains.

Eventually the track-bearing layers of central Texas were buried and converted to sedimentary rock. Millions of years later, as the Paluxy and other rivers carved their beds into those rocks, creating the modern landscape, the track-bearing strata became exposed in the riverbeds. Stream erosion scoured away the material that filled the fossilized dinosaur footprints, and the tracks became visible in the rock of the riverbeds, creating the illusion that the tracks had been made by creatures crossing the rivers.

By far the most common tracks in the Paluxy are the three-toed prints made by bipedal dinosaurs. Some of these footprints are as much as half a meter long. They usually have long, narrow toe marks, sometimes with indications of large, sharp claws at their tips. Tracks of this kind were undoubtedly made by large, dagger-toothed, meat-eating dinosaurs, technically known as theropods, beasts related to familiar dinosaurs like *Tyrannosaurus* and *Allosaurus*. Although one can seldom say for sure exactly which kind of dinosaur was responsible for a particular set of footprints, the likely maker of many theropod tracks in Texas was an allosaur named *Acrocanthosaurus*. Bones of this monster have been found in Early Cretaceous rocks at sites in Oklahoma and Texas, and the foot skeleton of this theropod has the right size and shape to have made the tracks.

Acrocanthosaurus was a big animal, reaching a length of 10 meters (33 feet) or more and a mass of 2 or 3 metric tons. Like other theropods, it generally walked on its toes, with its foot bones off the ground, like an added segment of the leg. Modern birds walk in the same way; because the knee joint of a bird is usually hidden by its body feathers, people frequently mistake a bird's ankle for its knee, giving rise to the common belief that the knees of birds point backward. Because theropods typically walked on their toes, the back edge of a theropod footprint usually was not made by a heel at the rear of the foot, but rather by tissues underneath joints between the foot bones proper and the first bones of the toes. We humans, in contrast, walk flat footed, with the full lengths of our foot bones (apart from the arch of the foot) in contact with the ground.

Theropod trails in the Paluxy River and elsewhere are of very narrow gauge; footprints made by the left and right feet are close to the midline of the trackway. This tells us that theropods walked with their feet directly beneath their bodies and did not sprawl like lizards. Left and right footprints follow each other in sequence, indicating that theropods walked and ran in a "striding" fashion, as do humans and big ground-living birds, and did not hop in the manner of sparrows and kangaroos. Footprints of bipedal dinosaurs usually point somewhat inward, toward the midline of the trail, showing that these animals were a bit pigeon-toed. Very seldom do marks of a dragging tail occur in theropod trackways; meat-eating dinosaurs usually moved like huge, animated seesaws, their tails carried well off the ground to balance the weight of their heads and bodies over their hips.

Students of dinosaur trackways use the term *pace* to describe the distance from a track made by one foot to the next print made by the opposite foot. A *stride* is the distance between two successive tracks made by the same foot. In trackways made by walking theropods, the stride is usually five to seven times the length of the individual footprints in the trail. The British zoologist R. McNeill Alexander derived a formula relating the speed of a track maker to its stride and the length of its footprints, based on observations of living animals. Alexander's formula suggests that bipedal dinosaurs walked at speeds of 5 to 10 kilometers (3 to 6 miles) per hour.

A few trackways, including some theropod trails at the ranch where I first worked on dinosaur footprints, have strides as much as 15 or 20 times the length of their constituent footprints. Unless these trackways were all made by unusually long-legged dinosaurs, at least some of the trackmakers must have been running, with estimated speeds of about 40 kilometers (25 miles) per hour—a pretty respectable clip for large animals.

The coastal muds in which the bipedal dinosaurs of Texas left their footprints seem to have been rather gooey and to have interacted with the trackmakers' feet in a plastic fashion. In many three-toed Paluxy River tracks, the toe marks extend farther forward within the rock than they do at its surface. When a dinosaur lifted its foot off the ancient tidal flat, mud surrounding the animal's toes collapsed inward, meeting above the middle of each toe mark to roof it over. Once exposed by the present-day river, these toe tunnels become filled with gravel and sand, and to get a true understanding of the shape of a three-toed track one often has to dig this material out. In a frequently murky river that is home to many water snakes, this procedure requires, if not caution, at least a certain fatalistic abandon.

Theropod tracks are impressively big, but they seem puny in comparison with the Paluxy's sauropod footprints. Well-preserved forefoot tracks of sauropods are rather rounded or even somewhat U-shaped depressions half a meter wide and long. The hind-foot tracks, however, are what really grab your attention. These immense prints can be a meter long, and sometimes even bigger. Hind-foot prints have three large, outward-directed claw marks along the front margin, and deep heel marks at the rear. Both forefoot and hindfoot impressions are often very deep, punched as much as 25 or 30 centimeters into what is now the riverbed.

Texas sauropod trails are of wider gauge than those of theropods, with left and right tracks some distance from the trackway midline. This doesn't mean that brontosaurs were sprawlers, however; they probably held their limbs straight down, in an elephantlike, columnar fashion. This conclusion is based in part on the way sauropod skeletons have to be put together in order for joints between bones to match up properly, but it is also reflected in the geometry of sauropod trails. Had the Paluxy River sauropods been sprawlers, the immense lengths of their limb bones would have put left and right sides of their trackways even farther apart than they are.

An earlier generation of paleontologists thought that brontosaurs were such heavy animals that their legs would have been hard pressed to support their weight, condemning them to a dreary life spent bobbing about in lakes and swamps like Brobdingnagian invalids in water baths. The Paluxy River tracks shot that notion down; the depth and clarity of the tracks were not what one would expect had they been made by half-floating animals. Nowadays, paleontologists generally believe sauropods to have been more terrestrial than aquatic in their habits, an interpretation fostered by reexamination of sauropod skeletal anatomy as much as by footprint evidence. Even so, I suspect that sauropods, like modern elephants, spent a fair bit of time in and around water.

The sauropod that made footprints at the Paluxy River and other sites in Texas was probably a creature known as *Pleurocoelus,* a dinosaur whose weight was perhaps ten times that of *Acrocanthosaurus.* Bones of *Pleurocoelus* have been found in rocks of the appropriate age in the Lone Star State, and its foot structure matches the tracks fairly well. A few years ago some colleagues of mine and I published a technical description of the sauropod prints and gave them the name *Brontopodus*, a name originally created, but never published, by R. T. Bird. Footprints are assigned different names than those applied to skeletons of the creatures that made the tracks because we can seldom be completely sure that we have a perfect match between track and trackmaker.

One of my odder experiences working in the Paluxy involved a beautiful trackway of a small sauropod; the trackway was uncovered by a flood that had removed an overlying thick bar of gravel. I was to lead a field trip of paleontologists to the Paluxy River that summer, but as my frequently bad luck would have it, another river rise reburied the trail before my trip. The head ranger at Dinosaur Valley State Park, Billy Paul Baker, and I decided to dig the trackway out so that my group could see it.

A month before the scheduled trip, Billy and I hauled out rocks and shoveled gravel for hour after backbreaking hour, putting on an impressive display of manual labor for park visitors while we became soaked with sweat and river water. Finally the tracks were once again uncovered. A gentleman who'd sat and smoked while watching the last half hour of our efforts bestirred himself to take a look at what we'd dug up. "You boys carve those things?" he asked.

To this day I wonder what kind of idiots or charlatans he thought we were, to have gone to the trouble first to manufacture fake footprints in the bedrock flooring the river, and then to work ourselves to exhaustion digging our bogus fossils from their gravelly shroud.

Exposing the fossilized tracks is only the first step in studying them. Individual footprints must be mapped, measured, and photographed, and which tracks go together in trackways must be determined. I seldom try to quarry dinosaur footprints from the rock in which they occur. It's much easier to make a copy of a track. The simplest way of doing this is to grease the track with a separator like petroleum jelly and then fill it with plaster of paris. Once the plaster hardens, a topographically reversed "negative" replica of the fossilized print can be lifted away.

Sometimes, however, plaster cannot be used for casting tracks. If there are undercuts in the print, such as the toe tunnels I've described, a plaster copy of the track would be hard to remove once it set up. One must instead use flexible casting materials like latex or silicone rubber. A latex cast can be reinforced with cheesecloth or burlap to make it stronger, and given a plaster or fiberglass backing so that it will be less floppy when it is removed from the original footprint.

In the late summer and early fall, water levels in the Paluxy drop to such a point that the river sometimes ceases to flow and becomes a series of still-water pools and intervening stretches of dry bed. So many footprints that are usually underwater are now exposed that this is the best time for making copies of footprints.

The trick to making a latex cast is to put the rubber paste down in thin layers, allowing the latex to dry completely before applying a new coat. The hotter the sun, the faster the rubber will cure, which makes the heat a bit more tolerable. I've made many casts on days when it was so hot that I was the only thing (barely) moving in the valley, apart from water snakes so intrigued by my presence that they repeatedly swam past to check me out.

Because a cast of a footprint is a permanent record of the track's three-dimensional shape that can be taken back to the laboratory for further study, including comparison with other dinosaur tracks and with the foot skeletons of dinosaurs that might have made the track, I make as many footprint casts as I can. I have frequently entered drugstores in small Texas towns to buy much of their supply of cheesecloth and Vaseline, prompting odd looks from the storekeepers and, no doubt, the suspicion that they were waiting on the hottest thing east of the Pecos.

The reason for going to so much effort to study and collect dinosaur footprints is that these fossils supplement information about dinosaurs obtained from working on their bones. Apart from what dinosaur trackways tell us about how the great reptiles walked and ran, footprint sites can also tell us what habitats dinosaurs frequented and whether particular kinds of dinosaurs were solitary or moved about in groups. For example, most of the Paluxy River sauropod trails head in the same direction, suggesting that they were made by a herd of these huge plant eaters passing through the area; such trails tend to show a common direction of travel at other sauropod tracksites as well.

A dinosaur could only die once, and so could submit but a single skeleton for possible fossilization. On the other hand, during its lifetime it might have made a huge number of footprints. This means that any individual dinosaur had a much better chance of leaving a record of its presence in tracks than as a skeleton. Fossil footprints can indicate that animals whose skeletal remains have not been found nonetheless lived in that region. In many rock units there are far more fossilized dinosaur footprints than bones, and so our understanding of dinosaur faunas in those areas depends on our ability to identify the tracks.

A major thrust of my research deals with just such matters. To what extent can we identify footprints as having been made by dinosaurs known from skeletal remains? How different can two three-toed footprints be and still have been made by the same kind of dinosaur? Or turning the question around, how similar can two such tracks be and yet have been made by different kinds of dinosaur? How might we distinguish footprints made by baby dinosaurs that would grow up to become monsters like *Acrocanthosaurus* from tracks made by adult theropods of small-bodied species?

These are tough questions, and more than one person has misidentified dinosaur tracks, often having been led astray by atypical prints made under unusual circumstances. Because of the somewhat plastic nature of the sediments in which they were impressed, three-toed tracks at Paluxy River sites seem to show distortions of shape more frequently than dinosaur footprints I have examined elsewhere, and this has contributed to a truly remarkable case of footprint misinterpretation.

To get to Dinosaur Valley State Park from Glen Rose, you turn off U.S. 67 west onto Farm Road 205. This road goes up and down and around hills in the scrabbly countryside, repeatedly crossing the Paluxy River. Immediately after one such river crossing and just before arriving at the park entrance, you see off to your right, across a large open field, a brown building and a house trailer that sports dinosaurs painted on its side. A sign at the entrance to the complex identifies this as the Creation Evidences Museum. A billboard near the entrance depicts a white-robed figure struggling to keep from falling off a cliff, his plight observed by dinosaurs in the background—the whole image reminiscent of publicity posters for the old Victor Mature movie *One Million B.C.* The billboard's caption announces that there is Evidence Here that man and dinosaurs co-existed. You might correctly suspect that there is, at the very least, an interesting story here.

In a way, it involves a search for field support that went bad. Dinosaur hunting can be an expensive endeavor, and R. T. Bird's boss, a paleontologist named Barnum Brown, had frequently financed his bone-collecting forays with money donated by the Sinclair Oil Company—in fact, Sinclair used a big green brontosaur as its corporate logo in Brown's honor. Brown repaid the company's generosity by calling his fieldwork Sinclair Dinosaur Expeditions, and so everybody benefited from this exercise in reciprocal back scratching.

Bird's first account of the Paluxy River footprints was published in *Natural History,* a magazine put out by the American Museum of Natural History. In addition to describing the sauropod and theropod prints he'd seen in the river, Bird also published a photograph of the carved man tracks he'd seen in the trading post in New Mexico, as well as a description of the alleged human tracks still in the riverbed. He did this in order to add, as he recalled years later, "a touch of mystery" to his article, hoping that this would amuse the head of Sinclair Oil and make him receptive to future requests for field money. It was, in effect, an exercise in what is today called "grantsmanship."

The unintended consequence of Bird's innocent little maneuver, however, was to bring the Paluxy River footprints to the attention of the right wing of Christian Fundamentalism, persons whose belief in the historical inerrancy of the Bible is accompanied by the conviction that the Creation account of the book of Genesis should be interpreted literally. If the first chapter of Scripture tells of the making of the universe in six days, then this means six literal days of twenty-four-hour duration—and forget any talk about evolution.

To R. T. Bird's great chagrin, his Paluxy River sites were pressed into the service of Biblical literalism: if human footprints were to be found in the same sedimentary rocks as dinosaur tracks, they could be used to challenge the scientific conclusion that people and dinosaurs had been separated in time by millions of years. A deluge of tracts, books, and even films described the "man tracks" of the Paluxy River, some hinting darkly that Bird and other paleontologists had conspired in efforts to cover up this embarrassment for "evolution science."

For many years there were more Creationists than mainstream scientists working along the Paluxy River. During one of my first trips to this area, I asked a passing teenager if he'd hold one end of a steel tape so that I could take some measurements of a sauropod trackway. He was obliging enough, but when we were finished he wanted to know what church group I was working for. The notion that I might represent a university hadn't occurred to him.

By the time I began working on Glen Rose tracksites, scientific Creationists had claimed to have found, in addition to dinosaur footprints, tracks of bears, saber-toothed cats, and both normal-sized and giant humans in the rocks of the Glen Rose formation. I was, naturally, very curious to see this stony footprint menagerie for myself.

I knew of two groups that had already made independent critical examinations of Paluxy River man-track claims. The first group was made up of field-workers from a Seventh-Day Adventist school, Loma Linda University, and the second was a party of scientists who called themselves the "Raiders of the Lost Tracks." Both groups had discovered that there was less than met the eye to most of the more exotic footprint claims. Apart from the handiwork of George Adams and his fellow footprint artisans—which had been accepted as genuine by many Creationists—

many of the alleged bear or human tracks turned out to be products of natural weathering and erosion of limestone, a fairly soft rock that is also quite susceptible to chemical dissolution. The bedrock of the Paluxy River erodes into a variety of interesting shapes. Some surfaces become lithic Rorschach blots—you could see UFO landing pads in them, were you so inclined.

Not all of the putative human footprints could be dismissed so easily, however, for some of the giant man tracks obviously were prints of some kind. R. T. Bird himself had speculated that they might have been made by a meat-eating dinosaur that happened to be wading in soupier mud than usual and sinking rather deeper into the mire than was generally the case. When the beast withdrew its foot, the fluid sediment might have collapsed around the toes, leaving an elongated impression that would bear a remarkable resemblance to a human foot. Soon after I began working along the Paluxy I saw many footprints that made me think that Bird's idea was on the right track.

I didn't know it then, but another dinosaur-footprint enthusiast had already come up with an explanation for the origin of giant man tracks that went well beyond Bird's interpretation. Glen Kuban had begun work along the Paluxy River in 1980, initially hoping to find evidence that the man tracks were genuine. He quickly decided otherwise and concluded that many dinosaur tracks with elongated heel marks had been made by dinosaurs that had, for reasons known only to themselves, abandoned their normal style of walking on their toes to walk flat footed, their foot bones touching the ground, across the muddy shorelines of ancient Texas. Although I was initially skeptical of Kuban's interpretation of how bipedal dinosaurs had been walking when they made the elongated tracks, I became more receptive to it when I came across descriptions of similar tracks, also seemingly made by flat-footed dinosaurs, at sites from other parts of the world.

In the mid-1980s, Kuban collaborated with Ronnie Hastings, one of the "Raiders," on the man-track controversy. The story was becoming more interesting than the simple matter of how dinosaurs had made elongated tracks. At the most significant Paluxy River man-tracksite, both alleged human tracks and more typical three-toed dinosaur footprints had now developed pronounced differences in color from the surrounding rock, due to weathering of the river's bedrock (such color differences had in fact been present even when the Creationists were working at the site, but at that time had been so muted as to be hard to see). In the process, the previously less distinct "man tracks" had developed three dinosaurlike toe prints at their front ends—toe prints with very little topographic relief.

Kuban and Hastings concluded that the color differences had originated in response to differences in weathering between rock formed from material that had filled the dinosaur tracks after they were made and rock formed from the slightly different sediments in which the trackmakers had originally walked. Their work received a good deal of media attention, and as a result most Creationists abandoned claims that human tracks occur along with dinosaur footprints in the Glen Rose formation.

Creationists are not the only persons enamored of the Paluxy River man tracks, as I learned when I gave a talk on dinosaur footprints at the Field Museum of Natural History in Chicago. In the course of my presentation I discussed the giant-man-track problem. Seated near the front row was a man with the most intense expression I have ever seen on the face of a human being. I just *knew* he was going to give me a rough time, and sure enough, during the question-and-answer period at the end of my lecture, his was the first hand to shoot up.

"Dr. Farlow," he asked, "what makes you so sure that the Ancient Astronauts were *not* fifteen feet tall?"

It was a question that I hadn't given much thought. I stammered through a reprise of my explanation of the man tracks, but it didn't satisfy him. He cornered me as I was leaving the stage, and he informed me that he had seen, at a Chicago art museum, Inca pottery on which were engraved images of Ancient Astronauts doing veterinary surgery on dinosaurs, and how did I explain that?!

I was clearly out of my league.

As if devotees of outsized antediluvians and extraterrestrial tourists weren't enough fun, there are still people who doubt the authenticity of any fossil footprints in the bedrock of the river. The same day that Billy Baker and I dug out the sauropod trackway and the man asked us whether we'd carved it, a woman watched me clean sediment out of one of the dinosaur's footprints.

"How do you know those things are real and that the river didn't just erode them into the rock?" she asked.

It wasn't a bad question, particularly considering some of the obtuse queries we'd already fielded that day. I pointed out the anatomical features of the track, showed her how these features could be seen in footprint after regularly spaced footprint in the dinosaur's trail, explained that it was highly unlikely that river scouring would reproduce features of such complexity again and again, and told her how well the footprints matched the structure of the dinosaur's foot skeleton. A good answer to a reasonable question, I thought smugly. But her follow-up came out of left field.

"Do you think people came from animals that lived in the water?"

I wondered how she got to that question from the one before it, and my spirits sank as I realized that once again I was going to have to deal with the perennially popular God vs. Darwin question. Sometimes I don't mind. I am a Christian myself, albeit not of the strictly Fundamentalist variety, and I often welcome the opportunity to explain my belief that a scientific outlook on the origin of the cosmos is not necessarily incompatible with the Christian faith.

But not this time. I was soaking wet, and now both physically and mentally exhausted. There'd been too many rocks to move and too many weird and even hostile questions for one day. I stole a glance at Billy, who gets questions like these all the time; he looked like he was about to explode with laughter. Maybe I could talk my way around it.

"No, not directly," I told the woman. She nodded and looked satisfied, and for a moment I thought I was off the hook, but then her expression abruptly changed.

"What do you mean, 'not directly'?" she asked sharply.

Oh, Lord, here we go, I thought. I said that, yes, I did think that the human species came from ancestors that, if traced species after species far enough back into the past, had once been marine animals.

And then the games began in earnest, back and forth. When our discussion finally ended, she informed me that after her death she was assured a place in heaven, but that I would be lucky to be reincarnated as a chimpanzee.

Nearly as murky as my prospects in the hereafter is the long-term fate of the Paluxy River itself. Texas law declares that all river waters belong to the state, and a permit from the Texas Water Commission is required before anyone can divert, store, or otherwise appropriate those waters. In 1985 the cities of Stephenville and Glen Rose, along with Somervell County (in which Dinosaur Valley State Park and Glen Rose are located), applied for such a permit, seeking permission to construct a dam on the Paluxy River. The applicants argued that constructing the proposed reservoir was necessary to ensure sufficient water supplies to meet projected near-term municipal and industrial demands.

Water rights are big business in the state of Texas, and until fairly recently the only considerations in such cases had been whether the requested water was in fact available or had already been committed to other municipalities or agricultural or other interests, and whether the proposed project was technically and economically feasible. In 1985, however, a new state law went into effect that made the Texas Parks and Wildlife Department (TPWD) an interested party in such proceedings.

Because the proposed reservoir's dam was to be built a mile or so upstream from Dinosaur Valley State Park, TPWD found itself an extraordinarily interested party. In part the department was concerned over the prospective loss of streamside habitat and the likely deterioration of recreational (swimming and fishing) opportunities within the park, but a greater worry was how the damming of the Paluxy would affect the dinosaur footprints that constitute the park's chief reason for existence.

Even though fossilized footprints are quite literally hard as rock, rock does weather and become eroded. The relatively soft carbonate rocks in which the Paluxy's dinosaur footprints occur are especially vulnerable to erosion—which is, after all, what has created many of the Creationists' alleged human "footprints." The river is constantly nibbling away at the bedrock over which it flows, and it is gradually erasing dinosaur footprints in the process. A really bad flood—and the Paluxy experiences quite a few such hell-raisers—can tear up the riverbed in a spectacular fashion.

That being the case, one might have expected the Parks and Wildlife Department enthusiastically to support the impounding of the Paluxy's waters. The reservoir could be used to prevent major floods of the kind that cause serious damage to dinosaur tracksites, and also to eliminate the less drastic but more common river rises that prevent park visitors from seeing the tracks.

Unfortunately, it isn't as simple as that. In presenting their petition, the two cities and Somervell County requested that release of water from the reservoir be limited to 2 cubic feet per second (cfs), arguing that this would be sufficient to satisfy downstream water rights. This didn't satisfy TPWD, however. Effects of this low water flow on natural habitats within the river valley aside, the department felt that a water release from the reservoir of as little as 2 cfs would leave many dinosaur footprints above water during the winter.

As rock freezes, water within its pore spaces expands as it changes from liquid water to ice, and this expansion can shatter the rock. When the Paluxy experiences periods of low water levels at the same time as freezing temperatures, the river's dinosaur tracks suffer severe damage. The Parks and Wildlife Department feared that the routine low water-release rates requested by the applicants would result in even faster destruction of the dinosaur prints than occurs now.

Moreover, occasional heavy floods, despite their negative effects on already exposed footprints, also uncover new tracks that were previously hidden by overlying rock layers or buried beneath gravel bars like the one Billy Baker and I dug into. So the river both gives and takes away footprints. TPWD feared that the effects of the proposed 2-cfs flow rate on the Paluxy's dinosaur tracks would be to eliminate the giving and accelerate the taking away.

Why not simply quarry the footprints from the riverbed, or at least make casts of them, and put them on display in a museum, away from the ravages of the river and the weather? If it were only up to me, that is what I would do, but the state of Texas does not have the money to pay for this kind of salvage, given the large number of fossil tracks that occur in the river. Furthermore, the park was designated a national natural landmark by the National Park Service in 1970. The tracks of Dinosaur Valley State Park are a significant part of our natural and scientific heritage, and the Park Service feels that there is value in letting the public see these fossils in place, in the bed of the river—the way R. T. Bird saw them. Removing the footprints from their natural setting might violate the conditions under which the park's designation as a national natural landmark was made and perhaps result in the designation's revocation.

In short, once all things are considered, the best way to ensure that the Paluxy's dinosaur footprints will continue to be available for public viewing—bad as that way is—seems to be to leave the river alone.

The case came before a hearing examiner for the Water Commission in the spring of 1986. TPWD requested that, if the reservoir had to be built, its rate of water release be increased from that proposed by the applicants to a range of 4 to 14 cfs, with a yearly average of 8 cfs. The hearing examiner recommended to the water commissioners that the reservoir's release rate be somewhat less than that, 4 to 12 cfs, with an annual average of just over 7 cfs.

In 1987 the water commissioners made their ruling—or rather, rulings. In May of that year they approved construction of the Paluxy River reservoir but specified that its water-release rate follow the recommendation of the hearing examiner. A month later, however, the commissioners voted to reduce the rate of water release to the 2 cfs initially requested by Stephenville, Glen Rose, and Somervell County.

The Texas Department of Parks and Wildlife then joined forces with landowners opposed to the construction of the reservoir to file suit against the Texas Water Commission, an interesting case of one state agency taking another to court. The plaintiffs claimed that the water commissioners had acted illegally. It was alleged that one of the commissioners had changed his mind about the merits of the case after the May 1987 ruling, in exchange for the support of the applicants' attorney and the similar support of a Somervell County judge in that commissioner's efforts to be reappointed to his job.

The commissioner, the lawyer, and the county judge naturally denied that anything unseemly had taken place, but in April of 1991 the state district judge hearing the suit ruled that the water commissioner had in fact been influenced by politics in voting the way he did. The judge has not yet rendered a verdict on the fate of the water permit, however, and a final decision on whether the reservoir will be built will probably not be made for some time.

For now, the river continues to flow freely, and I hope it will be permitted to do so indefinitely. The Paluxy River is, quite simply, a beautiful place. We have been quick to sacrifice such places for economic gain, and I would like to have this one left alone.

Whatever finally happens to the Paluxy, I will treasure the experiences I have had along and in it. I have walked and waded in the riverbed among the rock-frozen trails of prehistoric behemoths and in my mind's eye have seen the trackmakers sporting along the shore of a far more ancient body of water. I have shared quiet mornings with the trackmakers' avian kin and hot afternoons with reptiles whose ancestors survived whatever calamity it was that ended the reign of the dinosaurs. My time in the river has been time well spent.

THE SEARCH FOR EVE

MICHAEL H. BROWN

It was in 1987 that DNA came through with the type of promise we all expected but didn't dare ask: it was going to tell us about the origin of life. More specifically, it was going to tell us about the origin of *Homo sapiens*—modern man. Big, big stuff, on the order of proving (or disproving) the Big Bang theory.

DNA. Shoot for the internal stars. The universe within was now up for discovery. If DNA could tell proteins how to form specific hair and eye colors, if it could program for a boy or a girl, if it could serve as the microscopic computer tape for every feature we own and a few we don't—if it carried traits from generation to fledgling generation in secret code—then surely it contained the rough footage of our past.

Or was this biological alchemy? In 1987 we knew only this: a supremely confident cadre of geneticists and anthropologists from the University of California at Berkeley had penned a sensational article for the technical journal *Nature* that was entitled "Mitochondrial DNA and Human Evolution." Not exactly something that was going to cause queues at the newsstands, but under the bland and unrevealing heading (*Nature* is not known for its romantic juice) was a summary that caught the world of science by utter surprise. The scientists, Rebecca Cann, Mark Stoneking, and the late Allan Wilson, were proclaiming that a special type of DNA inherited only through the mother had allowed them to backtrack through time and establish when the human gene pool—modern man— first arose. "DNAs from 147 people, drawn from five geographic populations, have been analyzed . . . All these mitochondrial DNAs stem from one woman who is postulated to have lived about 200,000 years ago."

Stop the presses. Get Darwin on the line. These eggheads from Berkeley were claiming not just to have gained insights into evolution by using a new and wholly unfamiliar form of genetic analysis, but also to have located that most precious of all evolutionary players—Eve!

It was nearly like discovering the Ark. Using technology from re-combinant DNA engineering, Wilson, Cann, and Stoneking had narrowed down our point of origin, demonstrating that we all possessed the mi-tochondrial DNA of a single faceless but now chronologically distinct woman whom Wilson dubbed "Eve."

She lived somewhere in sub-Saharan Africa. Her clan had begun proliferating two thousand centuries ago, soon to conquer the globe. Her skin was obviously dark—which was going to be a considerable blow to white supremacists and all those who promoted European Neanderthals as our ancestors. She probably had one of the first skulls that was smooth and long and gracile—no more beetled brows. No doubt she could utter a few rudimentary sounds. She was callused and wore, if anything, a crude terra–cotta–colored cloak or a swathe of antelope hide. She lived where the rhino took dust baths, where the gazelles gathered, and where the giant buffalo roamed. She moved from acacia to acacia, foraging among the ambling giraffes and skittish, dust-kicking wildebeests, vervet mon-keys spilling from the branches—at night listening to the staccato rumor of hippos or a snapping branch indicating the presence of a leopard.

This was the savannah in 200,000 B.C. This was everybody's common ancestor.

According to these Berkeley scientists, Eve had probably risen from ape-man stock in a region that's now encompassed by Kenya, Botswana, and Tanzania. Modern humans derived, said Wilson, from Kung bushman lineages. We'd long suspected that Africa was the point of origin for apes and ape-men, but now scientists—hard scientists, geneticists!—were tell-ing us the Dark Continent was also the origin of anatomically modern man. In science this is known as a "eureka moment." It was comparable to evidence of comets killing off the dinosaurs or to the discovery of

Peking Man. Finally we knew where we came from and had a general notion of a great-great-great-grandmother whose DNA now finds itself in more than five billion human beings.

It was splashed across the cover of *Newsweek* on January 11, 1988. Didn't it touch everyone's cerebration? If DNA contained all the traits of our great-great-great-grandparents—and beyond—why could it not be deprogrammed to finally tell us where we originated and who our first mother or at least one common ancestor was?

Eve!

This is what intellectual dreams—and narcissism—are made of, the closest science comes to sexiness. It's not often that DNA theorists make both *Nature* and *Newsweek*. Eve, there all the while in our mitochondrial DNA. And why is that so implausible? If we can put men on the moon, bounce lasers off satellites, and separate protons—if we can drop smart bombs down Iraqi elevator shafts—then surely we can peer deep into the genetic material in each of our cells and decipher our cryptic history.

Let me explain how the theory and method worked in lay terms— that is, inevitably, in overgeneralizations and oversimplifications. DNA is a chemical substance—deoxyribonucleic acid—that mutates over long periods of time. Occasional tiny mistakes in reproducing itself cause it to vary from person to person. The mistakes can be averaged over time. Using various abstruse extrapolations, geneticists believe they can predict what percentage of DNA will mutate over x number of years, and thus, by counting up the differences, they can tell how long it has been since two people shared the same DNA and the same common ancestor. It will soon be possible, they speculate, to trace Cleopatra's genes in living humans.

Back to that even more famous and elusive female, Eve. What they found at Berkeley was that their 147 samples differed by an average of 0.57 percent. How long would it take mankind's mitochondrial DNA to differ by that crucial half a percent? Well, the Berkeley people believe that this kind of DNA (which is different from the more popular nuclear DNA and is located not in the nucleus but in mitochondria, the energy-producing organelles of the cell) mutates at a rate of 2 to 4 percent every million years. If you divide 2 and 4 percent by 0.57 percent you get 3.5 and 7, and if you divide a million years by 3.5 and 7 you get 142,500 to 285,000 years. The researchers concluded it would take that length of time to cause the differences in DNA that they noted in blood samples taken around the world. The median figure comes to 200,000 years ago—in the general time frame that fossil scientists like the Leakeys had hypothesized for the evolution of modern man from ape-man stock.

Why Africa? Because Cann, Stoneking, and Wilson had analyzed samples from the sub-Sahara and found them to be the most divergent. These samples had more mutations than the others, meaning that the process had been going on longer in the African population than among the Chinese or Europeans or Australians. When researchers used a computer to devise an ancestral tree, the deepest roots pointed to African origins.

"We see no indication of any deep lineage in Asia," Dr. Wilson later said.

"Africa emerges as our homeland," added Dr. Cann. "Africa has been a source of continual renewal for our species over the thousands of years of human existence and development."

Moreover, according to their theory, this band of the first modern *Homo sapiens* soon spread from Africa through the Middle East and replaced—outsurvived, and perhaps outright killed—the older Neanderthals (in Europe) and Peking man. There was no interbreeding. In this picture, those old cavemen popular science had once adopted as our ancestors were just dead ends on the genealogical tree.

The geneticists believed that DNA would replace fossils as the key tool of paleoanthropology. "Besides the likelihood that the most sought-after bones will never materialize, there is the problem of properly identifying bones that *are* found," noted Dr. Cann. "As molecular biologists we at least know that the genes available from present-day specimens came from some ancestor. In contrast, [fossil hunters] can never be certain that a given fossil has left any descendants." The dioramas in museums contained more fantasy, she added, than Spiderman's best escapades.

"When did the migrations from Africa take place?" the three scientists asked in their *Nature* paper. Well, the oldest cluster of DNA types to contain no African members was 90,000 to 180,000 years old, they noted, so the migration out of Africa may have occurred at about that time.

Let's say 135,000 years ago for the migration soon to define human destiny.

In all the years paleoanthropologists had been looking at stone tools and studying fossil skulls, in all the dreary years of unearthing tibia, no one had been able to come up with such a powerful theory about when and from where modern man migrated. "The Mother of Us All," blared a headline on the front page of the *San Francisco Chronicle*. Pulses were beating fast around the academic world. Geneticists were startled, biochemists cheered, fossil hunters worried that their art might no longer be relevant, and anthropologists were up in arms. The very use of the name *Eve*—which Wilson, who was famous for his histrionics, promoted—guaranteed an uproar not just in the confines of academia but among religious Fundamentalists as well. Were these highfalutin geneticists trying to say that the woman of Genesis had been located in the whir of a centrifuge? In mitochondrial genes?

No. Yes. Maybe. That was the Berkeley group's answer to a public that raised its collective head and stared toward the little laboratory spewing nitrogen and chlorides. The use of *Eve* clearly implied that besides discovering the general time and place of modern evolution, they had also shown that all of mankind derived from a single pair of *Homo sapiens*. Where was Adam? His type of DNA had not yet been analyzed.

Yet in reality none of the researchers were claiming that all humankind came from a single couple. Wilson's Eve was just a woman who existed at the time of the jump into modern physiology—*a*, not *the*, common ancestor.

The picture of human origins was taking the following shape: First there were ancient apes living as long as 32 million years ago; they led to man-apes known technically as *Australopithecus*. After *Australopithecus* came *Homo habilis*, the first man-ape thought to use stone tools, followed by a much more human-looking species called *Homo erectus*. *Erectus* was the coarse-looking, heavy-browed population that evolved into

Neanderthals in Europe and Peking man farther east. They were the first to use clubs, stone blades, and fire.

Erectus gave rise not only to Neanderthals but also to what anthropologists call "archaic *Homo sapiens,*" the African version of Neanderthals. They and they alone made the transition into anatomical modernity, according to the studies at Berkeley. They migrated up through the Sahara, across the Levant, and into southern Europe and western Asia, eventually supplanting the Neanderthals and descendants of Peking man without interbreeding with them. This raised the specter of prehistoric warfare. Did Eve's clan take over the planet by wiping out their more primitive competitors?

"Not credible," said C. Loring Brace of the University of Michigan. "Wacko," added one of his colleagues, an equally well-known anthropologist named Milford Wolpoff. "Interesting," commented the most famous of all fossil hunters, Richard Leakey, while his mother called use of the name Eve "nonsense." Any number of fossil hunters were disturbed at the Berkeley researchers' bold claims and infringement on their turf. "Bizarre," added Wolpoff. "Ludicrous!"

In short, the Eve theory has thrown the diverse worlds of anthropology, paleontology, and biochemistry into something of an uproar. Rarely in the recent history of science has there been a case that both captivated public attention and stimulated academic adrenaline to the extent of these DNA findings. Nor have there been many times when the shortcomings and biases of science have been quite so apparent. Let's see what kind of conclusions we can draw from the case of Paleolithic Eve.

First of all, the type of DNA the Berkeley group used is not the type of DNA that determines how we look, think, and act. That's nuclear DNA, which won't be fully analyzed for many years. Mitochondrial base-pairs, the type of DNA Berkeley used, are far simpler and found, obviously, in the mitochondria—bean-shaped organelles that serve as the energy-production centers of cells. Berkeley used mitochondrial DNA because it passes from generation to generation without recombination, which would scramble its base-pairs, making it near impossible to identify mutations. From this genetic point of view, mitochondrial DNA is like an hourglass in our cells.

Since the famous *Nature* paper, however, geneticists have indicated that the mitochondrial DNA all women currently possess did not actually have to come from a single female but more likely originated with a *group* of females who, at the time, had identical mitochondrial DNA.

Biochemist Wesley Brown, who originally taught the Berkeley group DNA technology, has also pointed out that mitochondrial DNA cannot really be relied upon as an evolutionary "clock." There is no evidence that it "ticks"—mutates—at a regular rate, and if it doesn't change at a regular rate, how can it accurately measure time?

There are dozens of other objections on genetic grounds. But some of the most compelling arguments come from the almost-forgotten fossil hunters. Anthropologists like Dr. Wolpoff argue that the European Neanderthal and the Asian *erectus* did not simply disappear without leaving us some of their genes, as the Eve theorists imply, but rather played an important part in our ancestry. His evidence? Certain *erectus* traits such as incisor shape and a round, flat face are still in evidence among modern Asians, while Europeans possess facial traits, especially in the shape of the nose, associated with Neanderthals. The cavemen of Europe and China may have given us as many traits as the cavemen from Africa. "Each region of the world has different features of what is primitive," explains Dr. Wolpoff, who is said to have analyzed more fossils than any other living morphologist. "The fact is that there are features like this that show continuity in China, and what they seem to show is that there are no invading African populations."

This opinion is echoed by the University of New Mexico's Eric Trinkaus, the most eminent authority on Neanderthals. "The idea that we had a very localized group of modern humans which emerged in Africa and then were the sole ancestors of all modern humanity is simply insupportable by the evidence," he argues. "All Berkeley may be documenting is the spread of *Homo erectus* (not modern *Homo sapiens*) out of

Africa. And if that is indeed the case, then the mitochondrial DNA, while very interesting, has no relevance at all as to the origin of modern humans."

Alas, while the Berkeley scientists have caused us to think anew, they have come up with no ironclad proof of the existence of Eve and are now being accused of sensationalism.

But that doesn't mean Eve never existed, and while Africa still remains a strong candidate for her homeland, new data from the Middle East suggests that the first *erectus* tribe to become fully modern may have lived in Israel. In a cave up a mountain just south of Nazareth—the French call it *la Grotte de Qafzeh*—fossil hunters have found a female skull that is anatomically modern and about 115,000 years old—the oldest anatomically modern fossil ever found, older than anything found thus far in the sub-Sahara or Europe. She's known as Qafzeh Hominid 9.

I took a trip to Israel to view the fossil, which is kept in a vault at the Rockefeller Museum in Jerusalem. It is dark from fire or iron oxide—a blackish rust color—but otherwise you couldn't tell it from a medical-school skull. If the face gives just a hint of the old jutting forehead, that's one of the few remnants of an archaic past, for she clearly had a rounded chin, a mouth like ours, and a smooth cranium. "She was only eighteen," said the curator, Joe Zias, with affection.

Eighteen. It seems so young for our 5,750th great-grandmother.

Who Was Neanderthal? Who Are We?

James Trefil

The men and women huddled around the fire, talking over the events of the day. Their callused hands, matted hair, and clothing of animal skins gave them a rough appearance, at least by modern standards. Anatomically, though, they were practically just like us—the same bones, the same teeth, the same bodies and brains. They were what scientists would later call Cro-Magnon man, or, to be more precise, anatomically modern man.

Standing up, one man dropped a flint tool into the fire. Perhaps he didn't notice, perhaps he marked the loss with a Cro-Magnon curse—it's not important. What is important is that after the flint fell into the fire, things started to happen. As the group settled down for a good night's sleep, the piece of flint warmed up. Deep within it, electrons that had been trapped in microscopic defects in the crystal structure were freed. They moved back into the main part of the crystal, giving off a small amount of light as they did so. I doubt if anyone saw the glow, and these people certainly wouldn't have known what to make of it if they had. But that soft glow is very important to us, as we shall see shortly.

The next morning the group went on its way. Their names, their lives, the very bones of their bodies passed from human knowledge, their only monument the piece of flint added to other Cro-Magnon debris near the ashes of their fire. But that tiny piece of rock was a telltale clue—as characteristic of Cro-Magnon times as a Coke bottle is of the twentieth century.

As the centuries passed, a slow change took place in the flint. Absorbing energy from cosmic rays and radioactive materials in surrounding rocks, electrons began to drift back into the defects from which the fire had driven them. Long after that fateful night, in a land that is now called Israel, scientists studying the evolution of *Homo sapiens* found the flint once more. Under the guidance of Hélène Valladas and her colleagues at the Center for the Study of Weak Radioactivity near Paris, it was heated again and the light it emitted was carefully measured. The amount of light given off by a rock in these circumstances depends on the number of electrons that have migrated back to the defects since the last heating and this, in turn, depends on how much time has passed since that heating occurred. By observing this phenomenon, which is called *thermoluminescence,* Valladas and her co-workers were able to calculate that the flint had been dropped into the fire some 90,000 years ago.

This result, published in 1988, may seem to be the outcome of one of those amiable but inconsequential scientific detective stories that crop up now and again. We may applaud the virtuosity of scientists who can ferret out this sort of knowledge, but it seems hard to work up enthusiasm for dating something so unimportant as the act of dropping a flint into a campfire. But in point of fact, that burned flint (and some nineteen others like it) forms a crucial link in a chain of logic aimed at answering one of the most profound questions we can ask about ourselves—"What is a human being?"

That these tiny bits of rock could be so important is the result of an old puzzle in the anthropology of the Middle East. We have known for a long time that the remains of both modern and Neanderthal man are dispersed across the Israeli landscape. Until Valladas and her colleagues made their measurements, however, we had no sure way of telling just when the region had been settled by each group. Was it a situation like the one in Europe, where the Neanderthals disappeared at the time modern men showed up? Were they there at the same time? It is in establishing timetables to answer these questions that the evidence of the burned

—

flints is crucial, for it seems to indicate that from 90,000 to 35,000 B.C. (the date of the disappearance of Neanderthal man), the two types of people lived side by side. There is no evidence in the fossils that any interbreeding between Neanderthal and modern man took place during this 55,000-year period, and this is the statement that catches the attention of anthropologists.

When a scientist approaches the question "What is a human being?" he or she does so by asking where humanity fits into the grand scheme developed by the Swedish biologist Carl Linnaeus (1707–78) that organizes all living things on our planet into categories. Human beings, for example, are members of the animal kingdom, the phylum of chordates, the subphylum of vertebrates, the order of mammals, the class of primates, the family of hominids, the genus *Homo,* and the species *sapiens.* At each step down this ladder of classification, we have fewer and fewer relatives. Squirrels, for example, are mammals but not primates. Snakes are vertebrates but not mammals, and so on.

When you see an animal described by two Latin names, it is the genus and the species to which these names refer. Giving these two names is roughly equivalent to giving the street name and number to locate a house. Thus, the familiar *Homo sapiens* ("man the wise") gives us the classification—the pigeonhole, if you will—into which to put humanity. But the problem faced by anthropologists is to consider what sorts of other beings are supposed to go into that pigeonhole with us. Do we share it with Neanderthal and other now extinct "humans," or do we occupy it in lonely splendor? In making this choice, we are, in effect, determining the characteristics that define what a human being is.

In classical biology, a species is defined to be an interbreeding population. If two animals cannot interbreed, they belong to different species. One way, then, of answering the question "What is a human being?" is to ask what sorts of organisms can interbreed with us. All existing human populations—Caucasian, Asian, and Negroid—can produce offspring with each other, so we are all members of the same species. Could Neanderthal man also be said to belong? The message of the burned flints, at least as it is being interpreted by some scientists, seems to answer this question in the negative. If the two populations did indeed live side by side without producing hybrid progeny, they were probably incapable of doing so and are therefore, by the classical definition, different species. This bears on a long-standing debate as to whether the Neanderthal is a subspecies of humanity (in which case he would be classified as *Homo sapiens neanderthalensis* and we would be *Homo sapiens sapiens*) or a different species (in which case he would be called *Homo neanderthalensis*). You will recognize that this debate, as picky and detailed as it may seem to be, is really about the question of what is and what isn't human.

It is altogether fitting that Neanderthal man should be coming back to the fore on this issue. The closest and most recent of our relatives, Neanderthal was also the first to be discovered. We have uncovered our family tree in "reverse," exploring the most recent branches first.

When workmen excavating limestone in a cave 60 feet above the Neander river near Düsseldorf uncovered some strange-looking bones in 1856, no one knew what to make of them. The partial skeleton was strange—the legs were short and strongly bowed, the skull had a low, sloping forehead with prominent ridges above the eyes, and the whole center of the face—nose, mouth, and jaw—seemed to be pulled forward. It certainly didn't look like any skeleton that had come from modern humans.

The basic problem faced by scientists confronted with this fossil was simple. At that time people believed that human beings stood at the apex of a creation made especially for them. We were thought to be different from other animals, isolated in splendor at the top of the tree of life. This was, after all, three years before Charles Darwin published *The Origin of the Species* and gave us a framework in terms of which we could think about human evolution. What were scientists in 1856 to make of these strange bits of fossilized skull and bone?

The dilemma was clear. If this were a primitive human, then we would have to admit that modern man was not uniquely created to occupy our position in nature. On the other hand, if the fossil were that of an ordinary human, there would have to be an explanation for its strange appearance.

My favorite attempt to reconcile the existence of Neanderthal man with conventional thinking was a theory put forward by Professor Franz Mayer of the University of Bonn. He argued that the skeleton was that of a Mongolian Cossack (with rickets) who had perished while participating in the pursuit of Napoleon's army across Germany in 1814. The idea was that the bowed legs were those of a lifelong horseman, the skeletal anomalies were due to rickets, and the pain of the disease had caused the Cossak to knit his brows, thereby producing the ridges. Professor Mayer did not explain how his rickety Cossack managed to scale a 60-foot cliff to get into the cave.

Even after the idea that man, like all other life forms, was the product of evolution gained widespread acceptance, the old ideas about our special place in nature retained their power. Other Neanderthal fossils were found, along with the first hints of still more primitive "humans." We found that Neanderthal man had thrived in Europe for a long period of time—from 150,000 to 35,000 years ago. But as much as we might accept the idea that our species was not specially created, the notion that we are somehow different from the rest of life found expression in a tendency to magnify the differences between "us" and "them." In this case, we made the gap between Neanderthal and modern man as large as possible.

Hence our present use of the term *Neanderthal* to denote a shambling, hulking brute. This picture, immortalized in the comic strip "Alley Oop," was actually based on the best scientific evidence available in the first half of the twentieth century. In 1913, the French anatomist Marcellin Boule presented a detailed study of "The Old Man of La Chapelle," a partial Neanderthal skeleton unearthed in 1908. According to Boule, Neanderthal man walked hunched over, with bent knees and a curved back, and was therefore very different from modern *Homo sapiens*. This brutish portrait of the Neanderthal's posture has generally been extended to his intelligence as well, although with less reason, since the Neanderthal brain is about 10% larger than ours.

It wasn't until 1956, at a symposium celebrating the hundredth anniversary of the original discovery, that anatomists William Strauss and A. J. E. Cave re-examined the La Chapelle skeleton and discovered that all the features that had led Boule to his conclusions about Neanderthal

—

posture resulted from the fact that this particular skeleton belonged to a man who suffered from advanced osteoarthritis. This led to our modern view of Neanderthal man as someone who looked more like Fred Flintstone than like Alley Oop. Neanderthal man was short (his average height was 5 feet, 4 inches) but very powerfully muscled, particularly in the neck and upper torso. His arms and legs were short but thick, and he walked erect. Dressed in a business suit, he could probably walk down a major city street without attracting much attention.

In addition to rehabilitating Neanderthal man at the physical level, we have found that he had quite a complex social structure as well. He buried his dead with ceremony, decorating the corpse and leaving materials in the grave. This practice suggests the existence of a belief in an afterlife and, hence, a religion. And as the physical condition of the Old Man of La Chapelle shows, Neanderthal man seems to have developed a mechanism for taking care of his old and infirm.

Many scientists believe that the Neanderthal also had a spoken language—something we normally associate only with human beings. In fact, the main debate seems to center not so much on whether Neanderthals had the capability of speech but on how complete that capability was. Some scientists, working with computer models of vocal tracts, have argued that the Neanderthal was incapable of forming the vowels *a, i,* and *u*. Others point out that chimpanzees can produce some of these vowels, even though the computer models say they can't.

The question of Neanderthal speech (or lack of it) got stirred up again in the summer of 1989, when a joint Israeli-French-American team announced the discovery of a fossilized Neanderthal hyoid bone in Israel. The hyoid is a little V-shaped bone that supports the base of the tongue and is connected to the larynx. Like most small bones, it is rarely found with fossil remains. In modern humans the hyoid is an important part of our speech apparatus, and apparently the Neanderthal hyoid is indistinguishable from ours. With this discovery, it seems to me, the weight of the evidence is shifting toward the view that they, like us, could talk to one another.

So the modern view of Neanderthal man is very different from the one that gave rise to the common usage of the term. All in all, he appears to have been a splendid fellow, not as different from us as we thought. Why, then, is there such a debate going on among paleontologists as to whether he and we should be put into the same Linnaean pigeonhole? Why do some scientists insist that he is our direct ancestor, while others hold that he is simply one more extinct side branch on the human family tree?

The question of our exact relation to Neanderthal man is an important one, because it goes to the root of the problem you encounter when you try to sort out the history of past life. If species are defined as inter-breeding populations, how do you classify animals that are extinct and can no longer mate with anyone? The problem is obvious when we deal with Neanderthal man, but it applies to the lowliest mollusk as well.

When scientists can't get access to the best evidence for answers to their questions, they see how far they can go with the evidence they have. For plants and animals that have become extinct, that evidence will be largely in the form of fossils. Fossils make it possible for us to examine remarkably detailed replicas of extinct life forms, but often, crucial parts of the organism simply aren't preserved. The inability to obtain fossils of soft tissue like the tongue and larynx, for example, is what fuels the debate over Neanderthal speech.

Another, perhaps more serious problem with the fossil record is that it is spotty—only 1 species in about 10,000 is preserved in stone, and even in a preserved species only a few individuals will be represented. Our entire knowledge of Neanderthal man, for example, comes from perhaps 12 skulls, 10 skeletons, 35 jawbones (with teeth), and fragments of bone from 300 to 400 individuals. Imagine trying to draw conclusions about all the human beings who have lived since the dawn of recorded history from such a skimpy sample! Yet that's exactly what paleontologists must do with Neanderthal remains.

In such a situation, subjective judgments about the importance of the thickness of a bone or a cusp on a tooth become crucial. "Sure, it's different," one scientist may say to another, "but is it *really* different?" After all, there is a large degree of variation within a single species—

think of a Saint Bernard and a Chihuahua. Deciding when two organisms are from different species on the basis of insufficient data demands judgment, experience, and not a little nerve.

The problem faced by the paleontologist, then, is that of sorting out the human family tree under the limitations imposed by the fossil record. And as these things go, the problems with Neanderthal man are relatively benign. He is, after all, a cousin only recently deceased, so his remains have not deteriorated as much as those of more distant relatives. In fact, one way of putting the question of whether Neanderthal is "human" or not into perspective is to look at the rest of our family tree (or at least at our present understanding of it) and see how he compares to other, less arguably "human" relatives.

One interpretation of the fossil data on our ancestors is that the last ancestor we shared with the modern great apes was a chimpanzeelike African creature called *Proconsul* that lived about 10 million years ago. The name has an unusual origin—in the 1930s, a trained Chimpanzee named Consul was popular in London vaudeville shows. When the skull of a very old apelike creature was discovered, the discoverers gave it the name *Proconsul* (before Consul) in a fit of whimsy. The first animal who may have shared the branch of the tree leading solely to modern humans was discovered in India. It walked upright, had a jaw similar to ours, and is called *Sivapithecus* (Siva's ape). Our knowledge of both of these transitional forms comes from a few skull and jaw fragments and the odd bone.

After *Sivapithecus,* we encounter the largest gap in the tree. From about 8 to 3.5 million years ago, we have no fossils and no evidence of development. Following this gap, though, is a major find—the fossil we call Lucy. She was discovered in 1974 by Donald Johanson, then of the Cleveland Museum of Natural History. A young female, Lucy stood 3 feet tall and weighed around 60 pounds. She walked upright, had a brain about a quarter the size of ours, and is the first known member of the hominid family. Johanson called her *Australopithecus afarensis* (southern ape from the Afar triangle in Ethiopia). She and the dozen other members of her kind that have since been found were among the earliest members of the genus *Australopithecus* (southern ape), distant cousins to mankind, and hence are sometimes called the first "humans." On the day the discovery was made, the paleontologists celebrated with a late-night party for which the background music was the Beatles' "Lucy in the Sky with Diamonds." Thus was the fossil of the earliest human named.

As time passed, other species of the "southern apes" appeared on the scene, lived for a time, and became extinct. Some of their lines disappeared 2 million years ago; some lasted almost another million years. One line, comprised of slender ape-men paleontologists call *Australopithecus africanus* (southern ape from Africa) apparently gave rise to a larger species that is the first member of the genus *Homo*. Dubbed *Homo habilis* (man the tool maker) because his remains are found with crude stone tools, he lived in East Africa from 2 to 1.5 million years ago. He was either ancestor or cousin to another human called *Homo erectus* (man the erect), whose remains are found throughout Africa and Asia. Most of the famous fossils you've heard of—Peking man and Java man, for example—are of this type. *H. erectus* had a brain about two-thirds the size of ours, which makes him the largest-brained of any of our relatives other than Neanderthal. This brain was probably instrumental in allowing *erectus* to develop another ability we associate with humans—the ability to use fire. *H. erectus* was around for a long time, from about 1.7 million to 500,000 years ago, at which point the early members of *Homo sapiens* appeared on the scene in Africa. Neanderthal, who appeared only 150,000 years ago, is a very recent member of the human brotherhood.

There are several features of this family tree that are important to note. One is that although today there is only one surviving member of our genus and family, this has not always been the case. It is quite possible that there have been times in the past when more than one member of the genus *Homo* shared the African plains with members of the genus *Australopithecus*—different species, different genus, but the same family. What would the meetings between bands of these different "humans" have been like?

As time went by, whether because of competition, direct aggression, or interbreeding, the other branches of our family tree died away, leaving us to inherit the earth. Thus, when Neanderthal man disappeared 35,000 years ago, it wasn't the first time that a flourishing branch of humanity had become extinct.

There are long spans of time during which the fossil record is too sparse to tell us what happened. Thus, we do not know whether *H. habilis* was the direct ancestor of *H. erectus*, the relative of some as-yet-unknown human, or an offshoot from the main branch. That such gaps in our knowledge should exist is not surprising, particularly when you consider that a few hundred fossils may be all we have to tell us the story of our kind over a period of a million years.

In the early days of the debate about Darwinism, there was much talk of the "missing link"—an intermediate step on the evolutionary ladder between man and ape. In the popular mind, the search for the missing link became synonymous with the verification of the theory of evolution. But we did not descend from the apes; both we and the apes descended from a common ancestor. There are not one but many "missing links"—a whole complex assortment of "humans" with whom we can claim kinship.

At what point along the branchings of our family tree do we stop and say, "Everything from this point on is a human being"? Moving forward from Lucy, we find beings that seem progressively more human. Lucy walked erect but had a small brain. *H. habilis* had a larger brain and tools; *H. erectus* had all this, a still-larger brain, and fire. For all of them, however, the question of whether or not they were human is easy to answer. No one would have difficulty in telling the difference between a 60-pound, furry, upright creature like Lucy and a modern teenager. But when we come to Neanderthal man, the differences aren't so striking and the question becomes more complex. Compared to the differences between us and Lucy, the differences between us and Neanderthal man are tiny. Unlike our more distant relations, Neanderthal man forces us to hone our thinking about the true nature of humanity.

To understand how modern scientists think about the Neanderthal, I talked to Chris Stringer, curator of fossil hominids at the British Museum of Natural History. Sitting in an office that looks down from the museum's magnificent Victorian building onto the streets of his native London, Stringer appears far too young to be a recognized world expert on anything as esoteric as Neanderthal man. But youthful or not, he is the main spokesman for the Neanderthal Is Different school of paleontology, a point of view that currently seems to be in the ascendant.

"Neanderthal doesn't really fit in with either *H. erectus* or anatomically modern man," he says. To make his point he pulls out a key and opens a large safe that takes up most of one wall of his office. Inside is a series of polished wooden boxes, each bearing a label designating its contents. Inside each box, cushioned in thick foam rubber, are the fossil remains of one of the members of our family tree. Because there are so few Neanderthal fossils around, each one has a name, and seeing those labels is like meeting a set of old friends. Pulling out the skull of a young Neanderthal woman found in Gibraltar, Stringer points to the inside of the skull. "We can tell the size of the brain from the skull," he says, "but that doesn't tell us how the brain was wired." He cautions against making too much of Neanderthal's large brain, but notes that the parietal lobes at the back of the brain, where visual signals are processed, were larger than one might expect. "Perhaps this means that they had very good eyesight," he suggests.

He also points out the unusual rounded pattern of wear on the front teeth. "They apparently held things in their mouths while they hacked with stone tools. Sometimes you see little chips in the teeth where they missed their targets. From the chips," he continues, "you can tell if the individual was right- or left-handed." I am always amazed at the details you can sometimes extract from fossils.

When you look closely at the Neanderthal skull, the contrast with modern man's is obvious, even to a theoretical physicist like me. But to Stringer, divergent anatomy is not the most striking difference between us and Neanderthal. "The real difference is in behavior, in technology," he says. His argument: from about 2 million to 35,000 years ago, the primary human tool was the hand ax, a piece of stone with a few chips taken off to make one edge sharp. This tool was used by everyone— *Homo erectus,* Neanderthal man, and early modern man. Then, 35,000 years ago, a new technology appeared. A piece of rock was used as a punch to knock flakes off another rock, and each flake could function as a separate tool. The new technology produced lots of tools, easily made. "Neanderthal just couldn't compete," says Stringer, "and he wasn't able to adapt fast enough. He lacked 'planning depth'—the ability to adapt quickly to changes in the environment. Like many other species of plants and animals, he was ecologically displaced."

So for Stringer, the difference between us and the Neanderthals lies in whatever it is in the human brain that allows us to produce innovative technology in response to the demands of our environment. I find this point of view fascinating, because it provides a link between the question of what makes human beings unique and another equally interesting question about the Neanderthal himself.

The disappearance of Neanderthals in Europe remains one of the great puzzles of science. For 115,000 years they lived happily in what is now Germany, France, Spain, and neighboring areas. Then, about 35,000 years ago, they vanished from the face of the earth. What happened to them? This puzzle is particularly intriguing because the period of decline of the Neanderthal coincides with the appearance of modern man in Europe.

So the two questions are inextricably linked. Neanderthal man's disappearance is clearly related to the way he interacted with our distant ancestors. What you imagine this interaction to have been depends on whether you regard Neanderthal as a subspecies of *Homo sapiens* or not, and this, in turn, depends on what you take to be the defining characteristics of our species. You cannot answer one of these questions without answering the other as well.

Over the last few decades our notion of what happened to Neanderthal man (and, consequently, of what we are) has undergone an interesting evolution. The theories, together with a short summary of their main points, are listed below:

Rape and Pillage

In the early 1960s, a popular theory was that Neanderthal man was a peaceful, if dull, sort of fellow who was slaughtered by invading tribes of modern men. This theory was part of a view of humanity brought to public attention in a series of books by Robert Ardrey, with *African Genesis* being the best known. In Ardrey's portrait human beings are characterized by aggression and find their most natural social organization in the hunting band. Neanderthal man, like the woolly mammoth, was simply another species that fell victim to the blood lust of *Homo sapiens*.

Love and Sex

More recently, the trend has been to describe many of the key turning points in human evolution in terms of the need to form families and maintain monogamous sexual relations. For example, proponents of this view have argued that upright posture makes face-to-face sex possible, thus strengthening the family bond. From this point of view, the characteristic trait of the human being is his capacity for peaceful cooperation with his fellows.

Consequently, the Love and Sex theory is that Neanderthals interbred with the invaders to form the present human race. There is, in fact, a whole genre of fiction devoted to exploring peaceful scenarios for the displacement of the Neanderthals by Cro-Magnons, the most familiar being Jean Auel's Earth's Children series. On a somewhat more serious level, the Finnish paleontologist Bjorn Kurten, in his novel *Dance of the Tiger,* blamed the demise of Neanderthals on the (fictional) willingness of Neanderthal women to nurture mixed-race babies because of their infantile features and refusal of Cro-Magnon women to do the same for infants showing the heavy Neanderthal facial characteristics.

A variation on this theme is expressed in a theory that, while always in the minority among paleontologists, has had distinguished defenders throughout the twentieth century. Championed today by C. Loring Brace at the University of Michigan, this view holds that the Neanderthal did not disappear in the sense we have been talking about, but was an intermediate evolutionary step between *Homo erectus* and modern man.

ECOLOGICAL DISPLACEMENT

This theory, advocated by Chris Stringer, is making some headway these days. It depends on *Homo sapiens* and Neanderthals having developed simultaneously, but in different regions—Neanderthals in Europe, *H. sapiens* in Africa. Only in a few areas, such as those caves in Israel, did the two species live side by side. When modern humans moved to Europe 35,000 years ago, Neanderthal man simply couldn't compete for resources and died out. "It wasn't a holocaust," Stringer insists, "just ecological replacement, a process that goes on all the time." In this view, we would say that what makes humans unique is not only their ability to innovate but their ability to use those innovations to exercise some control over their environment. This is a relatively new notion, and has not yet been applied to our thinking about how the rest of the family tree developed.

Looking at this history, I am struck by how closely our thinking about the fate of Neanderthal man is tied to our deeply felt notions about the nature of humanity. If you believe that the vicissitudes of evolution have made mankind irretrievably vicious and aggressive, you have Neanderthal man as victim. If you believe that humans are innately cooperative and driven by the need for sex and/or companionship, you have Neanderthal man as partner. If you believe that humans will be innovative and display little concern for the effects of that innovation on fellow creatures, you have Neanderthal man as innocent bystander. One anthropologist I talked to, Holly Smith of the University of Michigan, put it very well when she said, "When we look at Neanderthal, we are looking in a mirror. We see what we think we are."

In a way, these shifts in our ideas about Neanderthal man reflect shifts in outlook in the general society, particularly if you allow a fifteen- to twenty-year time lag for a student to work his or her way up to a position of scholarly prominence. Scientists in the late 1950s and early 1960s were in school in the aftermath of World War II. Is it any wonder that the unparalleled savagery of that event was mirrored in the Rape and Pillage theory? The scientists who put forward the Love and Sex theories were in school during the 1960s, and we all know what that decade was like. And the new wave of thinking characterized by Chris Stringer is being advanced by people who were in school during the 1970s, when the oil embargo and the onset of competition with the Japanese was the focus of attention, making a theory of ecological competition a natural outgrowth of their perspective on the world.

It is unusual for scientific theories to be quite so changeable and easily subject to social interpretation. To a large extent, this flexibility is due simply to lack of the kind of data that would constrain the imagination. Happily, after a hiatus of decades, new fossil discoveries are being made, and these discoveries will surely have a profound effect both on our conception of what a human being is and on what happened to Neanderthal.

But while waiting for new data to come in, and with tongue planted firmly in cheek, I am prepared to make a fearless prediction. Sometime around 2005, a respected young scientist at a prestigious institution will announce the discovery that the primary characteristic of humanity is the acquisition of material goods. We will then have the ultimate transformation—Neanderthal as Yuppie!

WHY IS IT LIKE SOMETHING TO BE ALIVE?

ROBERT WRIGHT

Most bats appear to lead pretty meaningless lives. They spend much of their time hanging upside down with their eyes closed, and even with their eyes open they can see almost nothing. To track down mice, mosquitoes, and other prey, they employ echolocation, emitting high-pitched squeals that bounce back into their large, ugly ears and are then relayed to some sort of neural synthesizer, which weaves the signals into a representation of reality. The system seems to work; Mexican free-tailed bats, in the course of a summer sojourn to the United States, fly hundreds of miles a day, from rock cleft to tree hollow to cave, and in the process dispose of roughly twenty thousand tons of Texas insects. Still, what kind of life is that?

This question—or at least a related one—has occurred to the philosopher Thomas Nagel. In 1974 he published a paper called "What Is It Like to Be a Bat?" This essay is noted for a number of things. One of the things it's not often noted for, but on which I'll dwell, is its salutary effect on the game of Animal Consciousness.

Animal Consciousness, long a favorite pastime of people who think about thought, consists of sitting around and speculating about which, if any, nonhuman animals have consciousness. One problem with the game, before Nagel's contribution, was figuring out what we mean by *consciousness*. When applied to humans, the term refers loosely to the subjective, intangible side of our selves; it is sometimes equated with such hazy concepts as *mind,* or with even hazier concepts such as *spirit* or *soul.* But what grounds are there for deciding whether those words apply to, say, a Mexican free-tailed bat?

Nagel avoided such loaded terminology altogether and more straightforwardly defined what it means for an organism to have "conscious experience." He wrote, "The essence of the belief that bats have experience is that there is something that it is like to be a bat." In other words, it doesn't matter what the answer is to the question "What is it like to be a bat?" What matters is that an answer exists. If there is anything at all that it is *like* to be a bat, bats can be said to have consciousness. If there isn't, we can file bats under Unconscious, alongside rocks, self-cleaning ovens, and all the other things we presume it is like nothing to be like.

With a single question, Nagel had made Animal Consciousness a game anyone can play. Try this one: Do chimpanzees have consciousness? A glance at their hominoid faces, their expressions of anger, fear, and sympathy, suggests that, yes, there is something that it is like to be a chimpanzee. What about dogs? Surely it is like something to be a dog. (Specifically, it is like being a deeply affectionate, overly dependent, blindly loyal, occasionally ashamed person with four legs, fur, and only a dim comprehension of the laws of the universe.) And cats? Yes, by definition it is like something to be smug. Rats? Even if all you've seen a rat do is get shocked in a college psychology lab, you probably surmised that, at that moment, to be that rat was to be in pain. And pain, we all agree, is something.

Further down the phylogenetic scale, the game gets trickier. Do cockroaches feel terror at the flick of a light switch? When some malicious child assaults an anthill and the whole colony shifts into high gear, has anxiety suddenly filled thousands of little psyches?

Of course, we can never know for *sure* that cats or rats, or any animals other than ourselves, have conscious experiences, but most of us intuitively feel that they do. And Nagel's definition of consciousness makes it even easier to feel this way.

Now making the rounds is a new version of Animal Consciousness—new, at least, as philosophical games go; it is about a half century old, and it didn't really trickle down to the level of late-night college dorm-room discussions until the 1970s. The premise of the game is that consciousness may extend not only beyond humans but beyond organic life. Someone may someday build a computer that is complex enough to be *aware* of its calculations, or to feel fear, anger, or sympathy. Being such a machine will be *like something*. You know—HAL in *2001*, Rosie on "The Jetsons," etc.

Computer Consciousness is no penny-ante game. At stake is the conviction that there is something special about being human. It doesn't undermine our self-esteem to think that dogs can feel hungry or sad much the way we do, because we know dogs can never get a joke, wonder if there's life after death, or have any of the various other highbrow conscious experiences that place our species in a category all its own. But computers, in principle, could do all these things. Not now, of course; the world's most awesome supercomputer is, at best, as complex as a bird's brain and, even at that, can't do all the things birds can do. But someday . . .

Those who contend that computers will eventually be privy to experiences traditionally reserved for humanity are a diverse group. One large contingent could be loosely labeled *scientific materialists*, or perhaps *determinists*, but for short we will call them the *mechanics*, because they believe that human beings (and all other forms of life) are like GMC pickup trucks—entirely explicable in terms of engineering. They maintain that every aspect of behavior, sensation, and thought is a product of the processing of information. And by information they mean *physical* information: hormones, synaptic firings, sound waves, light particles, and so on.

If the mechanics are correct in this belief, then presumably it is possible, by controlling the flow of information in a computer, to replicate human experience with precision. Why should the computer's electrons be any different from the brain's synaptic firings? Information is information, right? And besides, even if it isn't—even if for some reason synaptic firings are the only kind of information that will yield consciousness—that's no problem; in principle, someday, we could build a computer that runs on synaptic firings. So one way or another, with the right hardware, the right software, and enough time, we should be able to create computers flushed with pride, riddled with doubt, or alienated by the rapid pace of technological change.

The average American is not a mechanic. If pressed, he or she would probably describe the relationship between human consciousness (or mind or soul or spirit) and the human body much the way René Descartes did three and a half centuries ago: the "thinking substance," as Descartes called it, is in charge of the physical substance; the mind gives orders to the brain, which, in turn, instructs the body. Thus, when a boy says to his younger brother, "I slugged you because I felt like it, that's why," he is introducing him to Cartesian dualism. The word *I*, he is implicitly noting, can refer to either the subjective, insubstantial self or the objective, physical self, and the former governs the latter.

But if the subjective self rules the objective self, it doesn't do so with detachment. "My soul is not in my body as a pilot in a ship," wrote Descartes. The connection, he believed, is more intimate; the soul actually *feels* what the body feels and, indeed, can be temporarily commandeered by pain and other sensations deriving directly from physical stimuli. Thus, according to Descartes—and the average American—the causal connection between the physical and mental realms is two-way: Mind affects body and body affects mind.

The average mechanic would disagree. He (or she) would argue that consciousness has the same relationship to your brain that shadows in a shadow play have to the puppets producing them. There is a close correspondence between the two at any given time, but not because the shadow is influencing the puppets; rather, the causality moves entirely in the other direction: puppets determine shadows, and your brain determines how you feel. Thus the sense that some intangible aspect of yourself has power and somehow affects your neurological activity is all in your head. This is the illusion of free will under which we all labor every day.

To put the mechanics' point in technical terminology: consciousness is merely an *epiphenomenon* of physiological processes. It is affected by them but does not affect them.

Given this view of consciousness, it is easy to see why a mechanic wouldn't be threatened by the prospect that computers may someday be conscious. Shadow consciousness, after all, doesn't gum up the works; it may *feel* messy to the computer—all those sappy emotions burdening the binary spirit—but it won't upset the machine's smooth predictability. The electrons corresponding to it—the electrons that cast it as their shadow —will be running through the computer in perfect accordance with the programmer's dictates. Conscious computers, in this view, would never pull a HAL on us—stage a mutiny and start running our lives according to their whims. They wouldn't, after all, have whims.

This strikes me as a reasonable conception of consciousness. In fact (confession time) I'm basically a mechanic—a determinist, a scientific materialist. But there's one thing about the mechanics' conception of consciousness that really bothers me: if it is right, then we'll never know whether computers have consciousness. Shadow consciousness doesn't affect behavior, after all, and the computer's behavior is all we have to go on. The computer wouldn't say it had consciousness unless we had programmed it to say so, in which case we wouldn't know whether it was telling the truth. (Neither would it, unless, perhaps, it was.) So proof of the mechanics' view of consciousness—consciousness as epiphenome- non—can never come from computers.

Indeed, if the computer *did* suddenly step beyond the bounds of its program and start talking about how depressing Mondays are, it is the average American's view of consciousness that would be borne out: the causal link between the intangible and the tangible, between mind and body, would turn out to be two-way; the subjective realm would indeed be the source of forces that can unpredictably affect the objective realm.

Computer Consciousness, then, is a game we mechanics can only lose. If computers *do* someday evince a subjective life, our view of con- sciousness will have been undermined; if computers *don't* show signs of consciousness, this silence will be an annoying reminder that we can never know for sure that we are right about what consciousness is. And this state of perpetual ignorance will itself be a reminder—for those who need it: namely, us mechanics—that science has its limits, that the realm of subjective experience is in some sense beyond its reach. (Even mechanics, thanks to Thomas Nagel, have to acknowledge that this realm exists. We may avoid spooky words like *spirit,* and may even try to avoid the words *mind* and *consciousness,* but all of us concede that it is like something to be us.)

That consciousness is forever remote from the long arms of science is not, by itself, earthshaking news. Consciousness is inherently subjective, inherently private, and science by its nature deals with what is objective, publicly observable. Numerous people have noted this. They're right. But there's a second point floating around here that's seldom appreciated. Namely: not only can science not tell us if it's like something to be a computer—or even, for certain, if it's like something to be a chimp—science can't tell us *why* it's like something to be *us.* Science cannot explain the existence of human consciousness—not now, not tomorrow, not ever.

Science lives by the credo of the mechanics: everything about the structure and behavior of all organisms is explicable in sheerly physical terms, and therefore is amenable to scientific inquiry. This can-do spirit has gotten science where it is today, and I think it's the proper attitude of a scientist. But as we've seen, the flip side of the view that everything functional is physical is that what's not physical has no function. Our feelings, according to the credo of science, affect our behavior no more than a shadow affects a puppet. Well, if feelings have no function, then why are they here? A scientist—a mechanic—cannot answer that question. He can and should try to explain why evolution produced everything physical and observable about us—toenails, eardrums, brains, and all the intricate behavior brains govern. But if he's going to insist that consciousness doesn't do anything, he'll be hard pressed to explain why evolution created it.

In other words, a scientist can readily explain the evolution of the physiology that produces fear: adrenaline and various other tangible forms of information together got our ancestors to flee from saber-toothed tigers and thus were favored by natural selection. But why does that physiology produce, in addition to the fleeing, the feeling of fear? That's the tough question. If the feeling is truly superfluous, then there can be no evolutionary explanation of it.

If you still don't appreciate how little sense consciousness makes from a strictly scientific vantage point, it may help to take a whirlwind tour through our evolution.

Long, long ago, on a lifeless planet, some molecules started making copies of themselves—not because they magically sprang to life, but because their environment happened to induce some mundane chemical reaction that yielded more of them. The resulting copies of the molecules, for the same reason, did the same—and so on: copies of copies of copies. Copying errors were occasionally made, and any errors conducive to the survival and further replication of the resulting copies were preserved, whereas less helpful errors were not.

At some point a series of fortuitous copying errors led to the encasement of some of these self-replicating molecules in little cellular houses; and then at another point billions of years later, the houses were integrated into huge housing complexes—mobile housing complexes, no less, that could lumber around the surface of the planet. And these housing complexes processed much information—physical information, such as sound waves or light particles, that represented relevant aspects of the environment, such as food and predators. And the organs that processed this information were called *brains*. And the rest, as they say, is history.

The key thing is that at no point in this process—not even right up to the present—is there any discernible reason for subjective experience to enter the picture. There's no reason pain should be needed to get an animal to retract its hand from fire; this reaction, after all, is guided by a perfectly concrete flow of information, up the arm and back down it; the pain itself is thrown in at no extra charge. Similarly, there's no clear reason why the transmission of language has to involve anything other than physical things—sound waves, vibrating eardrums, synaptic firings, vocal-cord vibrations, more sound waves, etc. Computers can perform rudimentary language processing just by propelling electrons through silicon labyrinths, and we presume they'll someday handle complex language processing in the same fashion.

By the same token, there's no clear reason why, in order to process information about ourselves—about our appearances, about other people's opinions of us, etc.—we needed to become *self-conscious*. An Apple computer can process information about itself, and no one seems to think it's conscious, self- or otherwise. (It may be, of course; for all we know, it is like something to be a Macintosh—but this certainly doesn't seem *necessary*.)

This view of human evolution isn't just my personal view; it's the basic, if often unspoken, view within biology. Biologists try, with fairly relentless success, to explain all of evolution, including all its organic products and all of their behavior, by reference to sheerly physical factors. And yet, here we are, products of this sheerly physical evolution, infused with something undeniably ethereal: it is *like something* to be us. Who would have dreamed?

Now, all of this would amount only to a minor irony were it not for the single most interesting fact about consciousness: human life would have no meaning without it. Suppose for a second that machines could be made to do anything humans can do but did not possess consciousness. Upon sighting a comely female robot in the subway, a male robot would be compelled by his program to walk up to her and casually say "Come here often?" But he would feel no pulse of excitement upon first catching a glimpse of her, no wave of apprehension upon approaching her, and no embarrassment upon realizing that, because of a programming error, he had used his singles' bar line instead of his subway line ("Ride this train often?"). The physical correlates of these states of mind would exist; he would act excited and act apprehensive and act embarrassed. But he wouldn't really *be* these things. It wouldn't be like anything to be him.

If you stumbled onto a planet of such robots, there would be nothing immoral about murdering a few of them, because their lives would have no meaning anyway and would be incapable of acquiring any. They would possess no potential for pleasure or pain, for satisfaction or regret. This planet would have cohesive teams, yet no sense of camaraderie; stable marriages, but no love. No matter how human these machines might seem, their lives would have less significance than that of a Mexican free-tailed bat relishing the aftertaste of a mosquito, because the meaning of life lies somewhere in the fact that it is like something to be alive.

This fact gives new moment to the question of why consciousness exists, and it renders more frustrating the inherent inability of science to provide an answer. As long as we've concluded that life has meaning—compared, at least, to the amount of meaning any strict scientist should expect it to have—it would be nice to know why.

We can propose theories as to why, but to do so we have to step beyond the scientific frame of reference into the realm of metaphysics (which means, alas, that these theories cannot be tested, as scientific theories can be). For example, we can imagine some universal law of metaphysics dictating that when information is processed with a certain degree of sophistication, it is like something to be the information-processing system; complex information processors, in other words, must, in this universe, cast consciousness as their shadow. Or we can go further and posit the existence of a God, a God that (who?) had the choice of making life either meaningless or meaningful and opted for meaning. And, for that matter, we can combine these two theories and imagine a deistic God, a God that works (or worked and has since retired) only through laws of physics and metaphysics—a God that decreed meta-physical laws endowing information processors with consciousness, then decreed physical laws ensuring that complex information processors would evolve, and then, presumably, went fishing.

Note that these metaphysical theories don't contradict the credo of science; they just supplement it. They don't deny that consciousness is epiphenomenal; they just suggest why it exists, its epiphenomenality not-withstanding. Of course, you could also come up with metaphysical the-ories that *do* take issue with the credo of science, and thus with the mechanics. You could say, for example, that there's more to feelings than meets the eye, that they actually *do* something, that they *are* needed to animate certain types of physical systems (say, for example, animate ones). You could even start rhapsodizing about free will.

This last paragraph is further than I feel compelled to go. I will stick with the more modest metaphysical speculations, the ones that I can entertain without being stripped of my mechanic's credentials. Even these speculations will strike many of my fellow mechanics as weird. But it seems to me that when you're thinking about something as weird as consciousness, complementary weirdness is permissible.

Whatever the reason it's like something to be alive, the fact that it is should be savored and pondered liberally to relieve some of the despair that the modern scientific worldview can induce. It is true that evolution is a cruel process and has saddled us with all kinds of unfortunate baggage—greed, envy, hatred, ingrown toenails, etc. It is also true that the universe often seems to exhibit a disturbing indifference to our fates. But at least our fates matter to *us*—not just in the sense that we behave so as to avoid certain fates and to realize others (a thermostat can do that—and I'm fairly sure it's not like anything to be a thermostat), but in the sense that the fates we seek feel good and the fates we avoid feel bad. It is *like something* to be a complex product of evolution. And, so far as these products can tell, it didn't have to be that way.

THE MIND
OF A NEUROLOGIST

HAROLD KLAWANS

*"It wouldn't hurt to be nice to your aunt, would it?" she asked.
"That depends upon your threshold for pain," he replied.*
*—Conversation reputed to have taken
place between Mrs. Kaufman and
her son, George S.*

It was 6:45 and I was right on schedule. I'd been up working and sipping coffee for close to an hour and a half, and now I was in the shower. All I had to do was shave and get dressed, and I'd be out of the house by 7:00, on my way to the hospital for my real day's work. It's not that what I do early in the day isn't real work. I do my writing then, but it doesn't feel like work at all. I shampooed what little hair I have left, rinsed out the shampoo, turned off the water, opened the shower door, reached for the towel, began to dry myself off, stepped out of the shower stall with my left foot, as I had done a hundred or more times, swinging my right foot over the ledge—and missed! Pain. Excruciating pain exploded in my head! My right great toe had not swung over the ledge. I had somehow managed to smash it into that inch-and-a-half-high ledge. My broken right big toe. My big toe that was just beginning to heal. Why? Why this morning? Why not some other morning? Why did I have to be so damn clumsy when I already had a broken toe? Why?

The pain was beyond belief. It flooded my brain. My entire foot was aflame. My whole leg. Why did this morning have to be different from all other mornings?

On all other mornings my foot had glided safely above that ledge. On all those mornings when a stubbed toe would have meant only a stubbed toe. Not reinjury to a healing fracture. Why? Why when I had a broken toe?

Precisely, I suddenly realized, because I had that broken toe. Eureka! I felt more like Archimedes in a bathtub than a mere neurologist hopping on one foot outside a shower stall. I had not been clumsy. Not in the least. I was not a klutz. I had been suckered by my own brain. By some primitive defensive system of my brain that I'd never even heard of. It wasn't my fault. And as my toe throbbed and jolted away, I recognized that I had learned something about the brain, something that explained events that happen to everyone, usually more than once, but that remained unmentioned in any neurology book or journal I had ever read. Eureka. Land ho!

I had broken my toe three weeks earlier when I tripped coming down the stairs at 5:00 A.M., a result of not turning on any lights and of my own clumsiness. It was a fracture of the great, or big, toe on my right foot, right at the joint. I diligently taped it to my second toe and went on with my business, including a whirlwind professional trip to Europe that featured four different countries in six days and a lot of walking through airports and various other places. All of that, I'm certain, delayed the healing process, but now, three weeks after my original fall, my toe had finally been improving. I didn't have to be a physician to realize that all I had to do was act like a patient and rest the toe. Not only could I now stand without discomfort, I could walk without pain. And then the tragedy struck. Why? Because, simply put, the toe had not healed completely and I could not actually walk without pain. Painlessness had been an illusion, a conjurer's trick perpetrated by my own brain. True, I had been walking without consciously feeling any pain. But there is a difference between absence of consciously perceived pain and freedom from pain. A big difference. Just ask my toe.

If a tree falls in a forest and there is no one there to hear it, does that crash make any noise?

You know darn well that it does.

And does a broken toe cause pain each time you slam your weight down on it, even though you feel no pain?

Hell, yes!

Then why don't you feel the pain? Why hadn't I, as I stood in the shower that fateful morning?

It's called tolerance. You put on a tie in the morning; the collar feels tight around your neck. Half an hour later you can't feel it at all. Or you put on your bra and the straps pull on your shoulders. Not for long. In a few minutes you don't feel them. It's as if the straps have disappeared.

Why?

Tolerance. The steady, ongoing stimulus of the collar or the bra no longer registers consciously in the brain. Why not? Teleologically, it is because that unchanging stimulus serves no purpose. If it were otherwise, our consciousness would consist entirely of myriad unchanging messages, the feel of our clothing, our shoes, our socks, our pants, our underwear. If you are not thinking about it, do you feel your underwear? No. Yet your underpants are touching your skin and stimulating nerve endings. As a result of that stimulation, those nerves send messages toward the spinal cord and then up to the brain. The messages are coming in—all you have to do to feel your underpants on your buttocks is to think about them.

So why don't you feel them all the time? Your brain adapts to a fixed steady input and no longer "feels" it. Adaptation. *Tolerance*. That's the preferred neurologic term. Your brain develops tolerance. The stimulus no longer reaches consciousness.

What does that have to do with my stubbing my broken toe?

Pain is just one of a number of primary sensations the nerves send up to the brain. Pain. Position sense—where a part of the body is in space. Touch. Temperature—hot or cold. Pressure. One of many sensations. And each of them is subject to tolerance. The hot water in my shower that morning, like every morning, felt a lot hotter when I stepped

—

into the shower than when I got to the second chorus of "Our Love Is Here to Stay." I usually sing old Broadway show tunes in the shower: Gershwin, Porter, Rodgers and Hart. By the second song, I have to make the water hotter. Tolerance. The same tolerance that keeps me from being plagued all day long by that collar around my neck. And causes me to feel for my watch to know it's still there. And my wedding band.

The pain from my toe, my broken toe, had obviously not gone away. The toe was not healed. It couldn't heal that quickly. Not in a fifty-year-old. All I had to do was look at it. It was still enlarged. The soft tissue was swollen. The bone itself had thickened as new bone was being laid down. And if I tried to move the toe, I realized how far from back to normal it was. The joint was half-frozen, hardly capable of any movement at all. And any movement I forced it to make was painful. Not uncomfortable, but downright painful. If I palpated it, or gave it a firm push, the toe was tender. Palpation, pressure, caused discomfort that was on a continuum with frank pain.

If pushing down with my thumb caused pain, if standing up on my foot the first thing in the morning caused pain, didn't stomping around in the shower cause pain? Standing on one foot, my right foot, the foot with the broken toe, in order to wash my left foot, and thereby pushing down hard with my right big toe for balance—didn't that make my right toe and its sensory receptors, the raw nerve endings that feel and respond to pain, send pain messages to my brain?

It did. Obviously. It had to.

But I felt no pain as I stomped away in the shower. I went right on singing. Bernstein. Then Sondheim. "Ladies Who Lunch," I think. And I didn't feel the pain because my brain had adapted to the constant low-grade input from the raw nerve endings of my broken toe. My brain had become tolerant to the pain, as it had to the hot water.

But it wasn't just tolerance. It had to be more than that. Why?

Each of us knows precisely where each part of our body is at each and every moment. That is because each part of the body—or more precisely, the nerves of each part of the body, especially the nerves of the joints—send messages to the brain telling their location, their position in space. The brain keeps track of all of these unconsciously. Automatically. Reflexively. We do not become consciously aware of these facts unless we move a part of our body or ask ourselves a conscious question. If there is a preconscious part of our knowledge that can effortlessly be called into consciousness, this is it. When you move an arm, you know where it is—know consciously—but then you no longer know. Another example of the brain turning off those inputs it no longer needs. Diverting these sensations out of conscious awareness, out of mind. While all the time, the brain maintains a state of unconscious intolerance, awareness—better yet, constant vigilance. Both conscious tolerance and unconscious awareness are necessary for survival.

A few examples: each morning when I get into my car, I duck my head and miss the top of the door frame. I've never hit my head on the door frame of my own car. Not once. Have you? I don't think about hitting it. I don't worry about it. It's not something that can happen. Just like I don't think about smashing my foot into the ledge around our shower stall. I know where my head is without thinking about it. But if I put on a hat, that hat hits the door frame as I slide into my seat unless I direct my attention to it.

I'm writing at my desk. I'm completely absorbed. I reach up with my left hand and run my fingers through what remains of my hair. Once. Twice. Three times. I know what I want to write. I start writing. I write all first drafts longhand. Without thinking, I put my left elbow back down on the desk. Softly. A perfect landing. Lindbergh touching down outside of Paris at Le Bourget. I never hit the "crazy bone," the nerve running along my elbow and at risk for hitting the edge of any desk.

Why not?

Because I know where my elbow is in space, or more precisely, my brain knows. Unconsciously. It's called *unconscious proprioception,* the perception or knowledge of the position in space of each and every body part. Of my elbow as I rest it back on my desk, of my head as I slip into the car, of my right foot as I get out of the shower.

But I had smashed my right foot into the ledge. My right foot with its broken great, big toe, the toe that was now causing me such terrible pain. Ergo, I hadn't known where my right foot was in space. I hadn't known it consciously. Of course, I wasn't supposed to know it consciously. I was not thinking about it. I was too busy working my way through some Gershwin tune. The problem was that I didn't know my foot's location unconsciously. That was where my brain screwed up.

So why didn't I? Or more precisely, why didn't my brain know where the hell my toe was?

The answer was obvious, like all such simple truths. It was there to be seen if you just looked at it. All it took was getting hit by the right falling apple. It was a matter of tolerance, but a far different and far more pervasive form of tolerance than I had ever been taught about. Tolerance to feeling your underpants or brassiere or collar doesn't mean that you don't know where your buttocks, breasts, or neck are in space. Wearing a tight collar and a tie with a big knot does not increase the chances of my banging my forehead or scalp as I get into my car. But in the case of my toe, tolerance to pain had done just that. I had not only developed a conscious tolerance to pain, I had—or rather, my brain had—automatically developed tolerance to unconscious proprioception, the unconscious sense of where my right toe and right ankle were in space. The phenomenon had to include the right ankle, because you lift the ankle to get the foot and toe over that all important ledge.

My tolerance had crossed two separate borders. It had spread from one order of sensation (pain) to another (position sense, proprioception) and from consciousness to unconsciousness. As a neurologist and neuroscientist I hadn't known that could happen. As a typical human being I now knew that it did happen, and on reflection I realized that it always had and that everyone else knew that it did, even though those of us who study the nervous system as a way of life had yet to sanctify the observation. It has happened to all of us. You are working around the house, hammering away. You hit your thumb with the hammer. Accidents do happen. Your thumb hurts. It kills. Soon it feels better. Back to work. More hammering. Within no time, within half a dozen hammer blows, you've done it again, smashed the same thumb with the same hammer. Why? Because in dampening the original pain, the brain has also dampened its knowledge of where the pain is, of where your thumb is.

And *bang*.

You did it again.

Pain.

Worse pain than before.

Pain has a unique place among sensations. It is the most primitive. It may in fact be the primary, earliest, and purest of all sensations. This primitive form of sensation is an integral part of the nervous system serving to protect the organism. And it retains many of its primitive features, even in man. The nerve endings that feel pain are just that, the naked, thin, fine endings of the nerves. Other sensations require specialized receptors. Not pain. Just the nerves themselves. Most—in fact, all—other sensations reach consciousness in man in the cerebral cortex, the mantle of gray matter, of nerve cells on the surface of the brain. An area far more developed and sophisticated in man than in any other species. Destroy the cortex or anesthetize it, and the conscious recognition of most sensation is lost: hearing, vision, position, touch, smell, temperature. But not pain. Pain is felt unconsciously, deep in the brain, in the thalamus, which for all other sensations (all but smell) is a clearinghouse for input on its pathway to the cortex and conscious awareness. And of course, thalamic pain is devoid of all the trappings of cortical sophistication. The cortex is capable of a far greater degree of recognition of gradation and location. It can make complex judgments. The last joint of the great toe is bent up by three degrees. That's a cortical sensation. The thumb is being held just two millimeters away from the first finger, and between the thumb and the first finger there is a cool, round, hard, cylindrical object. It is the nail, two millimeters in diameter, being held between the thumb and the forefinger, as the hammer comes crashing onto the nail.

But the hammer misses. It hits the thumb instead. Instant havoc. Right on the thumbnail, of course. I should have hired a carpenter. A handyman. Someone without ten thumbs.

Throbbing pain.

Where?

My whole hand.

No, my whole thumb, up to my elbow.

Pain is, essentially, poorly localized. This is because it is a thalamic sensation. Only when I move or touch my smashed digit do I know precisely where the hammer made contact. Moving or touching my thumb adds cortical components and localizes the source of the pain.

Pain that in a few minutes recedes.

Slowly.

Surely.

Thank God for tolerance.

I start to work again.

And *bang*.

DISASTER.

The thalamus lacks sophistication. It has no judgment. It dampens pain, and as it does, it also blocks position sense, preventing that sensation from reaching consciousness and even aborting unconscious awareness. That is why I hit my thumb that second time and the third time, and why you have done the same thing. And why I hit my broken big toe on that ledge.

I finally knew why it had been on that morning that the accident had occurred. Why that morning had been different from all other mornings.

COMPUTERS
NEAR THE THRESHOLD?

MARTIN GARDNER

The notion that it is possible to construct intelligent machines out of nonorganic material is as old as Greek mythology. Vulcan, the lame god of fire, fabricated young women out of gold to assist him in his labors. He also made the bronze giant Talus, who guarded the island of Crete by running around it three times a day and heaving huge rocks at enemy ships. A single vein of ichor (the blood of the gods) ran from Talus's neck to his heels. He bled to death when he was wounded in the ankle or, according to another myth, when a brass pin in his heel was removed.

After the Industrial Revolution, with its wonderful machinery, writers began to speculate about the possibility that humans as well as gods could some day build intelligent machines. In chapters twenty-three, twenty-four, and twenty-five of his novel *Erewhon* (1872), Samuel Butler wrote about the coming of just such robots. Tiktok, one of the earliest mechanical men in fiction, was a wind-up copper person who made his first appearance in L. Frank Baum's *Ozma of Oz* (1907). He was manufactured in Ev, a land adjacent to Oz, by the firm of Smith and Tinker. A plate on his back said that the robot Thinks, Speaks, Acts, and Does Everything But Live.

After the computer revolution produced electronic calculating machines, with their curious resemblance to the electrical networks of a human brain, the possibility of constructing intelligent robots began to be taken seriously, especially by leaders of AI (artificial intelligence) research, and by a few fellow traveling philosophers. Hans Moravec directs a robot laboratory at Carnegie Mellon University; in his book *Mind Children* (1988), he predicted the appearance of robots with human intelligence before the end of the next fifty years. Both he and Frank Tipler, a Tulane University physicist, are convinced that computers will soon *exceed* human intelligence, making the human race superfluous. Computers will then take over the burden and adventure of colonizing the universe.

Here is a passage from *Erewhon* that could have been written seriously by Tipler or Moravec:

> There is no security against the ultimate development of mechanical consciousness, in the fact of machines possessing little consciousness now. A mollusc has not much consciousness. Reflect upon the extraordinary advance which machines have made during the last few hundred years, and note how slowly the animal and vegetable kingdoms are advancing. The more highly organized machines are creatures not so much of yesterday, as of the last five minutes, so to speak, in comparison with past time. Assume for the sake of argument that conscious beings have existed for some twenty million years: see what strides machines have made in the last thousand! May not the world last twenty million years longer? If so, what will they not in the end become?

Another passage, remarkably prophetic, from the same book:

> Do not let me be misunderstood as living in fear of any actually existing machine; there is probably no known machine which is more than a prototype of future mechanical life. The present machines are to the future as the early Saurians to man. The largest of them will probably greatly diminish in size.

Actually, these are not the narrator's words but sentences that Butler attributes, tongue firmly in cheek, to an Erewhonian professor. The professor's opinions prompt the Erewhonians to destroy all their machines before they surpass human intelligence and take over the world.

In *Mind Children* Moravec puts it this way:

> Today our machines are still simple creations, requiring the parental care and hovering attention of any newborn, hardly worthy of the word *intelligent*. But within the next century they will mature into entities as complex as ourselves, and eventually into something transcending everything we know—in whom we can take pride when they refer to themselves as our descendants.

The most powerful attack on such opinions, which have come to be called "strong AI," is Roger Penrose's recent best-seller *The Emperor's New Mind* (1989). Naturally, the book was vigorously lambasted by strong AIers. Because I wrote the book's foreword, I too have been denounced for my obtuseness. This essay is an effort to set down in more detail precisely what I believe about the possibility that computers will soon be able to converse with us in ways indistinguishable from the conversations of human beings.

First, I should make clear that I am not a vitalist who thinks there is a "ghost in the machine"—a soul distinct from the brain. I believe that the human mind, like the mind of any lower animal, is a function of a material lump of organic matter. Although I remain open to the Platonic possibility of a disembodied soul, as I am open to any metaphysical notion not logically contradictory, the evidence against it seems overwhelming. Strong arguments for a functional view of the mind are too familiar to need summarizing.

If a human has a nonmaterial soul, it is hard to see why the same should not be said of an amoeba, a plant, or even a pebble. A few panpsychic monists such as Charles Hartshorne actually do say this, but I consider it an absurd misuse of words, a "category mistake," to talk of a potato in a dark cellar as having what Butler called "a certain degree of cunning." Here are two panpsychic quotations from Butler's imaginary Erewhonian professor:

But who can say that the vapor engine has not a kind of consciousness? Where does consciousness begin, and where end? Who can draw the line? Who can draw any line? Is not everything interwoven with everything? Is not machinery linked with animal life in an infinite variety of ways?

Shall we say that the plant does not know what it is doing merely because it has no eyes, or ears, or brains? If we say that it acts mechanically, and mechanically only, shall we not be forced to admit that sundry other and apparently very deliberate actions are also mechanical? If it seems to us that the plant kills and eats a fly mechanically, may it not seem to the plant that a man must kill and eat a sheep mechanically?

I agree with Aristotle that the self is the "form" of the body, or in modern terminology, a pattern of the molecular structure of organic matter inside our skull. Of course the pattern is far more complex than the pattern of a vase or the Empire State Building.

Life did indeed evolve along continua, but there are spots (albeit with fuzzy edges) where wide chasms were crossed and new properties of matter emerged. The first great threshold was the emergence of life itself from lifeless compounds. And the last of the great thresholds, the greatest of them all, was the evolution of a brain with such properties as consciousness (self-awareness), expanded free will (with all its moral implications), a sense of right and wrong, a sense of humor, the power to communicate complicated ideas by speech and writing, and a raft of creative skills such as the abilities to compose poetry and music, paint pictures, discover significant mathematical theorems, and invent scientific theories capable of being tested. With the last skill came the awesome power to control the process of evolution and steer it in new directions, as well as the power to terminate the process.

It goes without saying that many of these human traits are possessed to a weak degree by animals. Chimpanzees seem to have a low-level awareness of themselves and an ability to make decisions. (*Free will* and *self-awareness,* by the way, are for me two names for the same phenomenon, like Einstein's principle of the equivalence of gravity and inertia, or, what amounts to the same thing, the equivalence of gravitational and inertial mass. I cannot imagine myself having free will without being self-aware, nor can I conceive of being self-aware without some degree of free will.) Monkeys have a feeble sense of humor; you can watch them play pranks on each other in a cage. Bower birds may have a dim sense of visual beauty. Apes can communicate with humans by signs, and so can a dog or cat. A chimp can make and test conjectures about how to get a banana from the ceiling if there are boxes lying around. And so on. That animals feel emotions of love and pain is undeniable. It is equally undeniable that a major gulf of some sort was crossed when humans evolved from bestial ancestors.

The question is not whether our human traits emerged as a function of an evolving brain, as I assume they did, but whether it will be easy or difficult, perhaps even impossible, to build a calculating machine complex enough to leap the same threshold. At this point we touch the central theme of Roger Penrose's brilliant, controversial book *The Emperor's New Mind*.

The difficulty in crossing the threshold, Penrose argues, is that we don't yet know enough about matter to know how to do it. Clearly we know far more about particles than did Democritus, but we are still a long way from understanding those particles. In standard theories, matter is made of leptons and quarks, and these particles are taken to be geometrical points, or at least *pointlike,* as physicists prefer to say. In recent superstring theories they are not points but inconceivably tiny loops. In either case, Newton's hard little pebbles that bounce against one another—the kind of matter Bishop Berkeley ridiculed as a "stupid, thoughtless somewhat"—have now totally dissolved. What is left is mathematics. Leptons and quarks, whether points or loops, are not made of anything. Their fields are just as ghostly. On the quantum level, to put it bluntly, there is nothing except mathematical patterns.

My readers know how impatient I am with some pragmatists, phenomenologists, and subjective idealists of various schools who heap scorn on the notion that mathematical structures are "out there" with a reality that is not mind dependent. For these thinkers, mathematical reality is located within human experience. Like Penrose and the overwhelming majority of eminent mathematicians past and present, I am a Platonist in the sense that I believe mathematical patterns are discovered, not invented. Of course, they are still invented, in a sense. Everything humans do and say is what humans do and say. Mathematics obviously is part of human culture, but to say so is to say something utterly trivial. The fact that only humans can talk or write about mathematics and laws of physics does not mean that it is useful to deny that mathematics and laws of physics are embedded in an enormous universe not made by us, but of which we are a part, and an inconceivably tiny part at that.

As I have said before, if two dinosaurs met two others in a forest clearing, there would have been four dinosaurs there—even though the beasts were too stupid to count and there were no humans around to watch. I believe that a large integer is prime before mathematicians prove it prime. I believe that the Andromeda galaxy had a spiral structure before humans arose on Earth to call it a spiral. As the noted Bell Labs mathematician Ronald Graham recently put it, mathematics is not only real, it is the *only* reality.

Some eccentric philosophers prefer to think that human minds alone are really real. There are even physicists, overwhelmed by the solipsistic tinges of QM (quantum mechanics), who like to talk the same way. But the human mind is made of molecules, which are in turn made of atoms, which are in turn made of electrons, protons, and neutrons. The protons and neutrons are made of quarks. What are quarks and electrons made of? Nothing except equations. Let's face it. You and I, at the lowest known level of our material bodies, are made of mathematics, pure mathematics, mathematics uncontaminated by anything else.

The most elegant theory of matter today is of course QM. Unfortunately, it is riddled with mysterious paradoxes. In recent years Einstein's EPR paradox (named with the initials of Einstein and two of his colleagues) has been the most debated. How can the measurement of one particle cause the emergence of a property on a correlated particle that can be millions of light-years away? It seems to happen either instantaneously or with a speed faster than light can travel between the particles. In the first case, the paradox violates the dogma that prohibits instant action at a distance. In the second case, it seems to violate relativity, which prohibits information from traveling faster than light.

None of the many proposed resolutions is satisfactory. The many-worlds interpretation of QM seems to get rid of the paradox, but there is an enormous price to be paid. One must posit billions upon billions of ever-proliferating parallel universes in which everything that can happen, does. Other efforts to solve the EPR paradox do no more than restate it in a different language.

For years Penrose has maintained, along with David Bohm, Paul Dirac, Erwin Schrödinger, and other great physicists, that QM is not the ultimate theory of fields and particles. This was Einstein's own view. Indeed, it was Einstein who first proposed the notorious EPR paradox in an effort to show that QM was incomplete. Working physicists, for the most part, never worry about such things. As long as QM works, and of course it works magnificently, they simply accept the fact that (as Richard Feynman liked to say) QM is "crazy." Don't try to understand how it works, Feynman warned his students, because nobody knows how it works. Should physicists leave it at that? No, insist Bohm and Penrose, because that tends to discourage research that may some day find that QM, like Newton's gravity, is only a good approximation of a deeper theory. Penrose himself is trying to go deeper, with a geometrical theory of particles and fields about which I am not competent to have an opinion.

Penrose contends, and I agree, that until we know more about matter on a level beyond QM, we will not understand how our minds can be a function of our gray matter. Until we know those deeper laws, we will not even come close to constructing a machine that can do everything our minds can do.

In Penrose's opinion, the great mistake behind the optimistic predictions of strong AIers is the assumption that machines made of wires and switches, operating with algorithmic software, can cross the great threshold. Let's look at this assumption more closely. We know from the work of Alan Turing and others that it is possible in principle to build computers out of any kind of equipment that transmits energy along channels, with switches to tell the energy where to go. You can build computers with networks of pipes that hold a flowing liquid. You can build them with rotating gears, with string and pulleys, with little balls that roll down inclines or slide along wires as on an abacus. Mechanical devices of these sorts have been constructed in the past. If you are interested, you can read about them in my *Logic Machines and Diagrams*.

Every machine, the philosopher-mathematician Charles Peirce once observed, is a logic machine in having aspects that model logic functions. The blades of an eggbeater rotate in one direction "if and only if" you turn the handle clockwise. They go the other way if and only if you turn the handle counterclockwise. An old mechanical typewriter is a jungle of binary logic relations. (Peirce, incidentally, was the first to show how a simple logic machine handling binary functions could be built with electrical currents and switches.) A few years ago a group of computer hackers constructed a machine out of Tinker Toys that played perfect ticktacktoe. There is no reason why, in theory, one could not build a Tinker Toy computer that could do everything a Cray computer can do or, indeed, what any super calculating machine of the future could do. Of course, it would have to be monstrously large and intolerably slow. Would its sluggishness dilute its consciousness? Could it still write a great novel, provided it had a few thousand years to do it?

Now, no one in his right mind would say that a Tinker Toy ticktacktoe machine "knows" it is playing ticktacktoe any more than a vacuum cleaner knows it is cleaning a rug or a lawn mower knows it is cutting grass. Sophisticated computer programs that now play Master chess differ from ticktacktoe programs only in the complexity of their algorithms. A computer with such a program is no more aware it is playing chess than an eggbeater is aware it is beating eggs.

Strong AIers believe that as computers of the sort we presently know how to construct keep growing in the complexity of their circuitry and software, they will eventually cross a threshold and become conscious of what they are doing. If one believes this, is not one forced to say that a Tinker Toy machine of comparable complexity, or even one made with rolling marbles, will cross the same threshold?

I admit that all this may someday be possible, but I agree with Penrose and such opponents of strong AI as the philosopher John Searle that it seems extremely unlikely. We know very little about how the brain of a fish or a bird works. We do not even know how memories are stored in the mind of an ant. It is true that electrical pulses are silently shifted about inside the skulls of animals, but this is done in a manner far from understood. What Penrose is telling us is that evolution, working on computers made of meat, crossed a threshold in a way that involves laws of physics not yet known. It could be that if and when those deeper laws are discovered and we know exactly how our brain does what it does, we will be able to construct a replica (perhaps made of nonorganic matter, perhaps requiring organic molecules) that will simulate a human mind. But to expect a calculating machine made with components of the sort now in use or imagined to cross the threshold seems hopelessly unwarranted.

What does a computer do? It twiddles symbols—symbols that are meaningless until we attach meanings to them. It twiddles them in blind obedience to syntactical rules provided by the software. But our minds do more than twiddle symbols. They also twiddle meanings of symbols. I can easily imagine a monstrous machine made of Tinker Toys that can play Grand Master chess, but I cannot imagine it will know it is playing chess. By the end of this century I expect a chess program to be able to defeat any grand master while playing under the usual time restraints. Even now chess programs can crush grand masters when moves must be made within a few seconds. I expect that powerful computers will steadily improve in their ability to do all sorts of extraordinary things, but these things will all be done by symbol twiddling. I do not believe that the complexity of their circuitry will push them across the magic threshold.

Endless novels, stories, plays, and even operas have been written about intelligent robots, and it is no accident, I suspect, that so many strong AIers were science-fiction buffs in their youth. My favorite novel about robotics is a little-known one by Lord Dunsany, inexplicably never published in the United States. Titled *The Last Revolution* (1951), it concerns a rebellion against humanity of superintelligent, self-reproducing machines, the sort of rebellion that Tipler and Moravec believe possible. The book's funniest scene occurs when the narrator plays chess with the first prototype. Its inventor, Ablard Pender, lives with his aunt Mary. He pretends to wind up his "gadget" with a key so as not to frighten her by letting her know it is alive.

It requires only a few moves for the narrator, whose Ruy López opening quickly takes a bizarre turn not in any chess manual, to realize he is playing not only against an intelligence superior to his own but against a mind aware of what it is doing.

When Pender's girlfriend, Alicia, first sees the crablike, four-legged monster and its eyes like a cockroach's, there is an intuitive flash on her face like forked lightning. She senses immediately that the thing is alive. A dog, frightened by something it too knows is living but that has no smell, howls and bites the iron. The thing tears the dog to pieces.

The brain of Pender's robot consists of fine wire that transmits electrical pulses. "Did you make it entirely yourself?" Alicia asks.

"Yes, of course," Pender replies. "Don't you like it?"

"Time," says Alicia, "will have to show that."

Quantum Liaisons

D A V I D H. F R E E D M A N

Though 99 out of 100 physicists will state without hesitation or qualification that quantum mechanics is the correct and complete theory underlying all that one could hope to understand about our universe, there is still the odd physicist of solid reputation who maintains something may be amiss with modern physics. Of these, some suspect quantum mechanics as currently expressed is marred by some sort of error or omission, though perhaps only a maddeningly tiny one. Others insist that quantum mechanics, though completely correct, has not yet been fully and accurately interpreted. Abner Shimony is among the former group, Yakir Aharonov among the latter.

In this age of increasing specialization, one isn't merely a physicist. One is a high-energy-particle physicist, or a condensed-matter physicist, or a low-temperature-fluid physicist. Abner Shimony has picked one of the more interesting niches: He is an experimental metaphysicist. Specifically, he is trying to prove that quantum mechanics, generally viewed as one of the most accurate and widely applied theories in the history of science, is wrong. Well, not exactly wrong; "in crisis" is Shimony's classification of the problem.

Ever since it burst onto the scene in the 1920s, quantum mechanics has been stupefyingly efficient at predicting the inner workings of everything from stars to digital watches. The theory is also a mortal enemy of common sense. At its heart is the pronouncement that the result of any observation or measurement is influenced by chance; even worse, it insists that when an object is not being observed it is at once everywhere and nowhere. Physicists are quick to claim that quantum mechanical weirdness applies only to submicroscopic specks of matter like atoms and quarks, and not to everyday, full-sized objects like socks, which apparently disappear for other reasons. But physicists, who are normally regarded as a tell-it-like-it-is kind of people, are not being entirely ingenuous in this matter. A glance out the window does indeed provide evidence that the world at large is protected from quantum-mechanical bizarreness, but the truth is that no one has ever been able to offer a convincing explanation as to why this might be so.

Einstein fumed at this indefinite, chancy view of nature and charged that quantum mechanics was at best an incomplete theory. But Niels Bohr, Einstein's nemesis in this regard, responded with the physics version of Don't Worry, Be Happy: if quantum mechanics doesn't jibe with our view of reality, he said, then our view of reality is wrong. Generations of physicists have been only too happy to treat Bohr's intellectual shrug as the last word on the question, conveniently allowing them to employ quantum mechanics every day in their laboratories without ever having to fret over the fact that the lab and everything in it supposedly melt into a Twilight Zone of indeterminate possibilities as soon as they turn out the lights and leave.

Over the past three decades, Shimony has made it increasingly difficult for physicists to sweep this discrepancy under the rug, publishing and speaking widely on the topic. "What bothers me is the conflict between quantum mechanics and our ordinary knowledge of the world around us," he explains. "Physics shouldn't just be about solving puzzles and making laboratory predictions. We should be able to come away from it with a convincing worldview." If such talk sounds a little philosophical for a physicist, that's because Shimony *is* a philosopher; he received a Ph.D. in inductive logic from Yale in 1953 and is a member of the philosophy department of Boston University. If quantum mechanics seems a bit outside the jurisdiction of philosophy, bear in mind that Shimony is a physicist, too. He earned a Ph.D. in physics from Princeton in 1962, and he is a member of BU's physics department. In fact, it was the carrot of a dual appointment that enticed him in 1968 to leave a tenured position as a philosophy professor at MIT for a then-untenured position at BU. "I'm particularly proud of that piece of folly," he says.

Shimony may be theoretical physics's Ralph Nader, but he brings remarkable grace and good humor to the role. He isn't a small man, yet his sky blue eyes, unruly gray hair, and tendency to giggle lend him an elfin quality. He speaks in long, enthusiastic rushes, a manner that is perhaps inspired in part by the calcification in his inner ears that has gradually left him all but deaf. (A high-powered hearing aid and some lipreading enable him to understand most comments made in close conversation, but he has to sprint up and down the aisles of his lecture classes, Oprah-style, to field questions.) He is also quick to downplay his impressive credentials in not one but two of the most intellectually intimidating pursuits known to man. "I'm an amphibian," he says. "I can drown equally well on land and in water."

Shimony's deep fascination with both the technical and the ethereal dates back to his upbringing in an orthodox Jewish home in Memphis, where one of his earliest memories is of watching in awe as his father, a Hebrew schoolteacher, miraculously transferred wine between containers with a siphon. His parents encouraged a strong interest in mathematics, and Shimony says the thing he remembers most clearly about his bar mitzvah is mastering a challenging calculus technique before leaving for the synagogue.

Upon entering Yale in 1945, Shimony immediately declared himself a physics major, but he quickly gave it up for philosophy and went on to get his Ph.D. It was only then that he was suddenly bitten by the physics bug once again, and after devoting all his free time while serving in the Korean War to reading physics textbooks, he embarked in 1956 on a physics Ph.D. program at Princeton. "It was the hardest work of my life," he says. "I was ill prepared and overworked. It got to the point where I was doing physics problems in my sleep—not as nightmares, but actually solving them correctly." He did, however, have the good fortune to draw the celebrated quantum-mechanical theorist Eugene Wigner as an adviser just as Wigner was rethinking his own long-standing adherence to Bohr's pooh-poohing of intuitive notions of reality. Shimony was immediately captivated by the issue—an issue which, after all, he was singularly well equipped to pursue.

Shimony was not the first to try to wring from quantum mechanics a less murky worldview than Bohr's. The best-known alternative, advanced in the 1950s by physicist Hugh Everett, posits that every possible outcome of any observation actually takes place, albeit each in a different world. But the idea of uncountable trillions of parallel realities springing up every microsecond held little appeal for Shimony. "As my friend John Stewart Bell pointed out," he recalls, referring to the recently deceased theoretician, "this would have terrible implications for ethics. No matter how high a road you took in this world, you'd be doing something very low in another one."

Shimony's own tactic for reconciling quantum mechanics with the ordinary world is a surprisingly technical one: he is searching for a mathematical modification to the basic quantum-mechanical equation that will explicitly limit the shenanigans of multiple realities to the submicroscopic world, freeing physicists from having simply to *assume* such a limit. He is also responsible for several experiments that have forced physicists to stand nose to nose with some of quantum mechanics' hardest-to-swallow consequences. The theory predicts, for example, that an event in one location can instantly affect an event located at a vast distance—a brutal violation of common sense and hundreds of years of physics orthodoxy. Shimony and associates were the first to demonstrate the phenomenon in the laboratory, by making nearly simultaneous measurements on particles headed off in opposite directions and observing an otherwise inexplicable correlation between the results.

Perhaps in part because of these experimental credentials, Shimony's work has been taken seriously by the physics community—a community that has generally taken the point of view that a philosophical objection to a physics theory should carry about as much weight as a religious one. Five decades ago, Einstein's brilliant but philosophically motivated challenge of quantum mechanics earned him the reputation of being over the hill. Today, Shimony's call for a reevaluation of quantum-mechanical reality has been echoed by some of physics's most respected names, including Roger Penrose, Murray Gell-Mann, and (as we shall see below) Yakir Aharonov. Shimony thinks this turnaround represents the swing of an exceedingly long pendulum. "We're just getting around to recovering the point of view of the seventeenth century, when physics wasn't separated from philosophy," he says. "Scientists like Galileo and Newton were concerned not only with how things work but also with how the world is constituted. In many ways, we're reestablishing liaisons."

The great majority of physicists sneer that Einstein was simply dead wrong when he argued God wouldn't play dice with the universe—that is, that the universe couldn't be held captive to the whim of probability. But Yakir Aharonov takes a slightly more charitable view of Einstein's stubborn skepticism. "All those years, he was just asking the wrong question," he says. "It's not 'Does God play dice with the universe?' The question is '*Why* would God play dice with the universe?'" In asking himself the latter question, Aharonov has discovered that quantum mechanics tells us a great deal more about reality than anyone had thought.

Aharonov, a self-assured, nearly arrogant fifty-eight-year-old Israeli with a heavy accent and wry grin, is prone to hold forth on his theory in rapid-fire fashion through a haze of cigar smoke and creative pronunciation (though it's only fair to point out that his grammar is nearly flawless, and he frequently offers to put his cigar out, if in the same way in which one might offer to put out one's eye). He likes to start his "spiel," as he calls it, with Einstein in the 1930s.

At that time, desperate to show that probability wasn't God's style, while facing a tidal wave of acceptance for quantum mechanics, the father of relativity appeared to strike pay dirt when he realized an unrecognized consequence of quantum mechanics: a measurement made of one particle can instantly affect a particle trillions of miles away. Since his own theory of relativity posited that nothing, not even information, can travel faster than light, Einstein believed he had sounded quantum mechanics's death knell—and good riddance, as far as he was concerned. But the great man had failed to reckon with physicists' affinity for quantum mechanics. Okay, they said, so two particles somehow communicate faster than the speed of light; there's still no way to *control* this communication—the measurements to either one of the particles will always produce what appear to be random results—so no one could use the phenomenon to send instantaneous signals across the universe. Thus relativity is not violated, and we'll live with it. Einstein grumbled about this state of affairs until his death in 1954.

But even as Einstein's life was drawing to a close, Aharonov, then a physics student, was raising his own questions about quantum mechanics. "There has always been a crisis of intuition in quantum mechanics," he says, "and physicists have dealt with this crisis by not thinking about it. It became doctrine for every teacher to say, 'Don't ask questions. Wait until you learn to use the theory, then ask your questions.' But once students learn to use the theory, they become satisfied and they never ask."

Aharonov wasn't so easily distracted. When he was ready to begin his doctoral thesis in 1957, he approached his adviser, the well-known theorist David Bohm, to say that he thought he could prove that an electron could be measurably affected by a magnetic field even when the electron was completely shielded from the field—an idea that ran completely counter not only to common sense but also to all conventional interpretations of quantum mechanics, which declared that such effects could never be observed. Bohm gave him the green light and some advice, and two years later the prediction of the *Aharonov–Bohm effect,* as the phenomenon is now known, rocked the physics world. The effect was tentatively confirmed by experiment in 1960, and then more convincingly in 1986. John Wheeler, one of the great figures in quantum mechanics, has said that "the Aharonov–Bohm effect shows there are still a few surprises left for us from quantum mechanics."

Aharonov was convinced a much bigger surprise was in store, one that would reshape the strained relationship between quantum mechanics and reality. He didn't get much encouragement. Even Nathan Rosenberg, one of the two other physicists who plotted with Einstein to discredit quantum mechanics in the 1930s, tried to wave Aharonov onto a different track. Recalls Aharonov, "He said to me, 'I spent all those years working on it, and it didn't get me anywhere. Don't waste your time.'" But by the early 1960s Aharonov already thought he smelled blood: He had stumbled on a way to reformulate quantum mechanics mathematically so that the theory seemed to take into account the future as well as the past of a particle in predicting the outcome of a measurement. After publishing the highly abstract work in 1965, he put it aside. "I couldn't see what the importance behind the idea was," he says. "Besides, the mathematics was too complicated, and that's a sure sign that the thinking behind it wasn't clear."

Then, in 1984, in mulling over Einstein's thought experiment in which a measurement of one particle instantly affects a second particle some vast distance away, Aharonov was drawn back to his old future-past idea. It had suddenly occurred to him that according to the special theory of relativity—which states that observers moving at different speeds won't agree on the simultaneity of two events at different locations—an observer whizzing by at high speed could find that the second particle registers an effect just *before* the measurement is made to the first particle. "To that observer," he says, "it would seem that the results of the measurement to the first particle have come from the future to influence the second particle." Could such a bizarre effect be demonstrated?

A certain catch-22 built into quantum mechanics appeared to eliminate that intriguing possibility. Common sense declares that performing an experiment on a particle in the present has an effect on experiments performed in the future (you can confirm this in your backyard with croquet balls), and for once, quantum mechanics concurs with common sense. Thus if you want to test the effect of the future on the present, you'd better not make any measurements now, lest you disturb the future before the future has a chance to disturb what you're doing now. According to the time-honored interpretation of quantum mechanics, if there is no measurement, then there is no reality. In other words, the future may have an effect on the past—but only if you don't try to find out what that effect is. "That makes the entire concept sound meaningless," concedes Aharonov. But as with his discovery of the Aharonov–

Bohm effect, he refused to accept the idea that nature would build in something interesting and not provide us with some window, no matter how well hidden, with which to check it out. "That's the difference between a philosopher and physicist," he says. "The physicist wants to think about a real world in which experiments can be done; otherwise you could argue forever about which ideas were right, and the answer would just be a matter of taste."

To get around the problem, Aharonov started questioning some basic assumptions, and in 1986 he came up with an idea for a new type of experiment that accomplished what quantum-mechanical theorists had been saying for seventy years could never be done. The technique, which Aharonov calls a *weak experiment,* involves taking a large number of particles and observing the "average" behavior of the entire group, instead of trying to observe each one's quantum-mechanical behavior individually. "When you take a compass reading, you're measuring the direction of the earth's magnetic field, but you're not interacting with every magnetic particle in the earth strongly enough to affect that particle," he explains. "We should be able to do the same thing with a quantum-mechanical measurement." Such a "weak measurement" might then provide a clue as to what most of the individual particles are doing without disturbing them enough to affect the results of a later, more conventional experiment.

Aharonov began to employ weak experiments to probe the quantum-mechanical netherworld between measurements in a variety of exotic situations—all of them lavishly constructed, in the style of Einstein, in his imagination. "The advantage to being a theoretical physicist," he says, "is that you never have to worry about the cost of a thought experiment." What he found was that in most situations a weak measurement doesn't really tell you any more than no measurement does. Unless, that is, you happen to pick the right starting and ending points for the experiment, in which case the weak measurement proves everything Aharonov had hoped, and more.

Take, for example, the situation in which a weak measuring device is hooked up to a group of particles in such a way as to measure the group's total *spin,* a property of electrons and other particles analogous to the spinning of a top. Normally, the needle on the weak device's spin gauge will be nothing but a blur, reflecting the fact that in the absence of a strong, or definite, measurement, the particles take advantage of quantum mechanics' moratorium on reality for the unobserved. While a particle is in this limbo, it actually exists in a combination of all the many different realities that are available to it; it isn't until a strong measurement is made that the particle chooses from among them. In the case of the group of spin particles, many of these combined realities are associated with very large amounts of spin, balanced by an equal number of realities associated with small amounts of spin.

But Aharonov found that the situation would be quite different if the experiment were to start with the particles having their spins all in the same direction—spin direction being analogous to the direction in which the tip of a spinning top points—and end with the particles having their spins once again all in the same direction, except pointing nearly the opposite way. (Starting off this way isn't a tall order, because there are devices that can select particles according to their spin direction; but a group of a mere 15 particles would end up all in the opposite direction far less than 1 out of 100,000 times.) These starting and ending conditions have an extraordinary effect on the limbo state in between. Because these conditions are so restrictive and improbable, the particles lose much of their freedom of choice, in much the same way that an Olympic diver would lose much of his freedom of acrobatic choice if he were forced to leave the diving board in a certain contorted position and enter the water in another, very different one.

In losing their freedom of choice, the particles also give up most of the various realities that go into making up the limbo state, since the limbo state is now hemmed in by the unusual past and future measurements. In particular, the limbo state loses all the realities that were associated with small amounts of spin, because these small-spin realities aren't consistent with a group whose spins are all in the same direction. As a consequence, there is a runaway conglomeration of large-spin realities in the limbo state, and the weak measuring device clearly registers

—

a much larger total spin than the combined spins of the individual particles. "In this situation, all the low-spin states cancel each other out, and you end up with an impossible spin," explains Aharonov. "It acts as a kind of spin-amplifying device." In fact, he adds, if the spin measurements were tied to any type of force—say, for example, to the operation of a magnet—then this strange effect could serve to amplify that force. As weird as all this sounds, the spin-amplification effect was experimentally confirmed in 1990 at Rice University, and Aharonov believes it will lead to a practical force-amplification device in the not-too-distant future, perhaps one that could detect phenomenally tiny magnetic fields.

In 1989, after having worked out the theory behind the amplification effect, Aharonov got to thinking, What if a set-up like the one he had proposed were designed to amplify, not spin, not even force, but *time?* It was a truly Aharonovian musing—out of left field, extremely intriguing, and utterly impossible to test.

Or was it? Turning once again to Einstein for inspiration, Aharonov recognized that a constantly expanding and contracting balloon would slightly alter the gravitational field—and hence, according to general relativity, the flow of time—for anyone stationed inside the balloon. By linking the control for the balloon's expansion to a set of spin particles, one could thus hope very occasionally to find that the balloon's occupant would experience a leap in time, in either direction, that would be virtually unlimited in magnitude. It would be an odd sort of time machine, indeed: the time traveler could find himself "time translated" forward into a pile of dust, or backward into the atoms from which he was constituted.

Aharonov was amused by his discovery of a time machine, but he was more concerned with addressing a problem raised by his theory: If the results of an experiment can go back and affect the past, then why can't one simply arrange to run experiments whose results will reach back and change the results of an earlier experiment in such a way as to create a contradiction between the present and the past? This is just a technical version of the paradoxical murder-your-grandmother problem: if you go back in time and kill your grandmother, then you were never born, and so you never killed your grandmother, and thus you *were* born, and so on. That's a paradoxical situation physicists call *causality violation* and refuse to abide.

But even if the present can affect the past, causality violation isn't a problem as long as you can't control what happens in the present. And Aharonov realized that, thanks to quantum-mechanical randomness, you can't. "For any individual particle, I can, with an experiment, change its past," explains Aharonov, "but only in a way that's hidden behind uncertainty."

In other words, because quantum mechanics dictates that any experiment will always be subject to an element of randomness, the experimenter simply doesn't have the freedom to determine an experiment's outcome completely. That means you can't reliably rig an experiment to produce a result that will influence the past in such a way as to violate causality. And therein, from his point of view, lies Aharonov's real triumph: in showing how quantum-mechanical uncertainty acts as a sort of anti-causality-violation Scotchgard for the fabric of reality, he has given nature an excuse for intimately incorporating randomness in its structure.

Aharonov insists he knew there was a good reason all along. "Einstein thought quantum mechanics was ugly," he says. "But it is a lovely theory when you listen to what it is trying to tell you."

WALKING WITH TED
OR, A PHYSICIST CONTEMPLATES NATURE

HANS CHRISTIAN VON BAEYER

Every morning I take my dog, Ted, for a walk down to Lake Matoaka. When he was young, the exercise was one of those obligations that come with family life, but then it became a habit, and now it's an inflexible routine, as much part of my day as a morning cup of coffee. Ted is a pretty little fellow with the classic features, the luminous coat, and the billowing tail of a golden retriever, but his head only reaches up to my knees. I guess somewhere in his tangled heritage a growth gene got lost.

Ted's nose has a peculiar mottled pink appearance, and on top of his head there is a minute white spot you'd never find if I didn't point it out. He doesn't obey very well, and sometimes insists on rolling in some nasty-smelling indefinable substance he has dug up, which absolutely infuriates me, but the bond that binds Ted and me together is tougher than steel.

In the graveled parking lot across the street from my house, I remove his leash and he takes off in a brief burst of exuberant energy, but since he leads a pretty free life at home, he doesn't feel compelled to keep up this celebration of liberation for very long. Not unless something out of the ordinary catches his attention, such as our killdeer. Every spring a black-and-white-striped killdeer, strutting as if on stilts, builds her nest in the middle of the parking lot. Sheer stupidity, if you ask me, but I was glad to note that this year she had had the wit to choose a spot between the concrete ties at the ends of two parking spaces, so she had at least a modicum of protection. A killdeer nest is so inconspicuous that some-times you can't find it even when you know where to look. It consists of nothing more elaborate than a ring of pebbles around a shallow inden-tation in the dirt, and it holds four eggs that are covered with gravel-colored splotches. From the point of view of camouflage, a killdeer nest is an impressive demonstration of nature's efficiency: utter simplicity lead-ing to superior effectiveness.

I know that the nest has reappeared when the mother tries to distract me away from it by screeching piteously while dragging a faked broken wing through the dust. For me this maneuver achieves the opposite of the intended effect, but it works on Ted—not that he could ever find the eggs without my help. When mother killdeer goes into her routine, Ted charges her with ferocious gusto, but I suspect that neither party really believes in the seriousness of the attack. The killdeer simply flaps her wings a little faster and easily outdistances her adversary, while Ted's experience has long ago taught him the pointlessness of the act. Respon-sible dog that he is, though, he does what is expected of him.

After we turn into the path down to the lake and I duck under the barrier that keeps out cars, Ted tends to dawdle behind. There is a lot of sniffing to do, and quite a bit of ritual marking of the territory, so I leave him to his tasks and stroll ahead. It's downhill here, and the lake is still hidden behind the trees. In the spring, when the new foliage is just uncurling from every twig, and the ground is covered with May apples impatiently unfurling their umbrellas before they have finished pushing them through last year's soggy leaves, and the sun's golden rays slant down from behind me onto the dewdrops that sparkle like diamonds in the spiderwebs strung from branch to branch, and the smell of the warm air is full of vague promise, and the only sounds I hear are the first tentative morning chirps of robins and chickadees over Ted's muffled rustling—at times like that I feel a sense of peace with the world that comes close to spiritual reverie.

All around me life is new and bountiful. Wherever I turn I sense a restless pushing and growing, an invisible pulsating biochemical rhythm in living cells that is multiplied and amplified to the level at which it manifests itself in the motions of the forest's plants and creatures. Nature is stretching like a cat in the morning, and I'm glad that Ted and I are part of it.

A new leaf attached to a little twig falls down in front of me, probably bitten off by an overambitious squirrel. As my eyes trace its graceful descent, my mind, attuned to its motion by three decades of teaching physics, instinctively reviews the wondrous interplay of cause and effect unfolding before my eyes. According to the law of universal gravitation, the Earth below my feet, stretching all the way to our antipodal point somewhere in the Indian Ocean off the coast of Australia, conspires to pull that twig down in the direction of the center of the Earth. Every particle of matter in that vast globe that supports me stretches its unseen tentacles up to snare the little branch. I wonder if my brother Carl in Saskatoon knows that he's just helped to attract a twig in Williamsburg? I'm sure Ted doesn't sense his role in the microcosmic drama above his head, but he, too, is making his contribution. In fact, since he is so much closer, and since gravity diminishes with the square of the distance, his pull is vastly more powerful than Carl's. In principle, though, the entire world is involved.

By Newton's law of gravity, the twig experiences a downward force. By the same gentleman's law of motion, the twig responds by falling down in the direction of the force. But that's not the end of the story. I have taught Newton's laws for such a long time that they have become part of me. They inform the way I experience the world, and they help to shape my reactions to what I sense. So on my walk to Lake Matoaka with Ted, those laws come with me.

In particular, there is the law of action and reaction that so impressed young Thomas Jefferson when he learned it here in Williamsburg that he later incorporated it in the Constitution in the form of the principle of the balance of powers. The law of action and reaction guarantees that if the Earth pulls down on my little twig, the twig must pull up on the Earth with the same force. The gravitational attraction between two objects acts like a spring stretched between them, and, unlike an amorous attraction between two people, always tugs equally on both of them. So the Earth must respond by coming up to meet the twig.

"Poppycock!" I hear you exclaim. "The Earth is vast and solid and heavier than all get-out, and no twig is going to pull it around. Besides, the law of action and reaction applies only to two objects that interact with each other, like two billiard balls, or two houses of the legislature, or some other pair of weighty bodies—not to a single thing, like a twig, that is influenced by an external force." But you are wrong. The Earth does move.

One of the most valuable lessons of physics is that you learn to specify what the logicians call your *universe of discourse;* in plain English, that you become conscious of whom or what you are talking about. As long as I consider only the twig and the force that acts on it, then it is true that the law of action and reaction doesn't apply. But if I enlarge my perspective and think not only of the twig but also of Ted, and me, and my brother Carl, and Mount Everest, and the whole Earth down to our watery antipodes, then, suddenly, there really are two objects—the twig and the rest of the world. If I were to draw them on a blackboard I would indicate them by two dots, one to represent the twig and the other, the center of the Earth. And those two dots, which in reality are about four thousand miles apart, attract each other with equal and opposite forces.

The force on the Earth is precisely as strong as the force on the twig, but Newton's law of motion also specifies that heavy objects react more sluggishly than light ones. The property of the Earth that keeps it from jumping up every time a twig falls down is aptly named: it is called *inertia.* It's exactly the same tendency that keeps you from bounding out of bed in the morning, especially if you're tired or a little on the portly side. So, although the Earth moves up toward the twig, its great mass and imposing inertia hold it to such an inconceivably slow pace that the twig has landed on the ground and aborted the whole process long before the motion of the ground becomes perceptible.

But still, the reaction of the Earth is real, in principle. Quantitatively it doesn't amount to much, but qualitatively it must occur. As I walk down toward the lake with Ted padding patiently behind, I enlarge my universe of discourse and imagine not just me and my dog, but the two of us in relation to the entire Earth. And every time a leaf falls, or a twig drops, or a cricket hops down to a lower perch, every time I jump off the trunk of a fallen tree, the Earth recoils a little. The realization that our home planet is not a fixed, immovable stage upon which the drama of life is played out but a fellow actor, albeit considerably more bulky than the rest of us—that understanding enhances and deepens my relation to the natural world around me.

A squirrel crosses the path up ahead, and Ted takes off like a shot. But he doesn't even get halfway there before his natural inertia takes over and he slows down to a walk. I think all the squirrels in the woods know him by now, and they love to tease him. Years ago, when he was just a puppy and we first started these jaunts, he used to get pretty close to the squirrels and bark fiercely, and maybe even scare them a little. But after what I figure to be over two thousand trips and an astronomical number of threatening sorties, he has never once caught a single squirrel. I'm sure Ted knows by now that this is a game he is doomed to lose.

Just before we reach the lake, we come to a clearing. A few early buttercups have struggled through the sand and now stand up awkwardly on gangly stems, turning their bright little faces to the Sun like freshly washed teenagers. A turtle shuffles by in search of morning groceries. Ted doesn't even notice. The lake glitters through the trees. The earth's ponderous turning carries me slowly in the direction of our nearest star; time passes.

Overhead a white cloud hangs motionless in the blue sky. In school I learned about that color, and later I explained its secret to my students. Why is the sky blue? The proverbial question that is supposed to stump Dad and send him scurrying to the encyclopedia was answered a hundred years ago by Lord Rayleigh, who proved that molecules scatter blue colors much more effectively than red colors. When sunlight enters the atmosphere, it consists of a mixture of red and yellow and blue light. The yellow and red proceed unhindered, but the blue light bounces sideways and ricochets around the sky until it happens to be sent into my eye by an encounter with a molecule located somewhere in the atmospheric bubble above me. In the evening, when a beam of sunlight must traverse an extraordinarily long stretch of air, there is even more opportunity for blue to be removed from it than during the day, so the beam, robbed of even its last vestige of blue, looks bloody red. Thus Rayleigh achieved the efficiency science strives for: he explained two things, the blue sky and the red sunset, by means of a single theory.

Standing now near the shore of the lake, with Ted lying at my feet taking a not-quite-well-deserved rest from his moderate exertions, I look at the sky and understand its color. But as I ponder it, it seems to me that the answer could be phrased in different terms. Perhaps the very question is wrong. For what, after all, is the sky? Isn't it everything that is above me, out to infinity? Isn't it space?

Space, as we see for ourselves at night, is black. Except for the occasional star or planet, there is nothing out there to produce or reflect light. So when we look up in the daytime, we are not looking into space at all. What we see, rather, is the air in the foreground against an utterly black background. And, as it turns out, air is blue. When we look at nearby objects such as our hands, we look through a layer of air that is too thin to show its color—like a thin puddle of tea on a parquet floor —so we subconsciously assume that air is invisible or even forget about its existence altogether; but air is a material substance like any other, a bit thinner, to be sure, yet endowed with weight, and motion, and color. Just think of the wind and all the damage it can do: air is a real thing.

Painters, who are trained to look more carefully at things than the rest of us, have known for a long time that air is blue. You can tell if you notice that landscapes become more and more blue with distance. The dozen or so peaks and mountain ranges throughout the world that have names like Blue Mountain and Blue Ridge bear witness to the fact. Roman painters used the convention that increasingly blue hues signify greater background distance. Michelangelo wrote at length about the blue color of air.

What it all adds up to is that when I look up from the edge of Lake

Matoaka, I am not looking at the sky: I am looking at the air above me, and it happens to be blue. What I see is profoundly reassuring. Our old planet is not floating naked in the black, forbidding void of outer space. When we move about on its surface we are swimming safely in its protective envelope of blue amniotic fluid.

Ted never looks up at the sky—I don't suppose dogs have any reason to worry about its color. When we get to the dock, he takes a routine turn around it to check it out. It is rarely occupied. Occasionally a majestic blue heron stands watch there and upon Ted's intrusive yelping laboriously takes off like an overloaded bomber during World War II. The bird lands just as heavily on a stump sticking out of the water on the other side of the lake, huffily draws himself up to his full height, and stares at us with what I imagine to be an angry, unblinking eye. "And the same to you, too!" I call to him, because, after all, he doesn't own the dock.

From the edge of the path I pick up a stone and playfully fling it into the lake. The stone rises gracefully through the morning air, tips over, and ends its symmetrical trajectory with a plop. Like countless baseballs, footballs, and basketballs, like lumps of lava hurled out of bubbling volcanoes when the Earth was young and drops of water splashed up by oceans till the end of time, my missile traces a mathematical curve through space. We physicists find beauty in the timeless perfection of that motion. It doesn't matter whether the thrower is angry or happy, American or Australian, young or old. It makes no difference whether the day is clear, like today, or cloudy, or whether it is night or day. Neither the color of the projectile, nor its shape, nor its weight have the slightest influence on that curve. It represents a rare glimpse of the absolute in this chaotic world of ours. It is a truth, insignificant though it may be to the large scheme of things; truths are hard to come by, so we hang on to it. In the arc of that rock I see the motivation for pursuing my profession.

Galileo first derived the shape of the trajectory and found it to be the figure that Apollonius of Perga over two millennia ago called a *parabola*. "Ignoring air resistance, cannon balls move along parabolas," we learn in school. But when your universe of discourse has been broadened, the way mine had that morning, you find out that even this elementary-school wisdom, which rolls off the tongue with the greatest of ease, is only an approximation. The truth is more intriguing.

Imagine the stone as a point and the Earth all shriveled up and shrunk down to another point four thousand miles below your feet. This is how Newton, who was born in the year Galileo died, imagined it. If you simply release the stone, it drops straight down toward the center of the Earth like the little twig. But if you give the stone a push, the way I did when I hurled it over the lake, it follows a curved path through space. And from this enlarged perspective you see that the path is not a parabola at all. A parabola is an open figure with two legs that stretch off to infinity and never come back together, which is not the stone's route. Instead, the relation of the stone to the Earth is exactly the same as that of a comet to the Sun, and we know the shape of the comet's path: it is an ellipse.

The true figure of the path of the stone is a skinny ellipse, an oval that is about four thousand miles long and only a few miles wide at its widest, with the Earth's center just inside the lower tip and the stone on the edge of the upper end of the oval. Of course, the stone cannot follow that entire trip, because after just a few seconds it falls into the water—but that, to the physicist, is an inessential detail. The shape of its path through the air, before it sinks into Lake Matoaka, is the shape of the upper end of that almost unimaginably skinny ellipse.

The stone traces out before my eyes the trajectory of Comet Halley—but not the part that we see when Halley races in hot fury around the Sun—no, it imitates the other part that we never see, when Halley almost coasts to a stop thirty-eight years later and hundreds of millions of miles from the Sun and starts on its return journey, a thirty-eight-year fall toward the Sun. I have always wanted to be there when the great comet, far out in the dark cold of outer space, moving almost impercep-tibly slowly, comes to the apex of its odyssey and begins its long haul back home to the warmth of the Sun. But I don't need to go that far away. A stone tossed over Lake Matoaka mimics precisely what I would see out there beyond the orbit of Pluto. It is the business of physics to find unity in the diversity of natural phenomena—and to discover analogies between the inaccessible realms of the universe and the immediate world of human experience.

Ted and I start our own return journey up the path to go home. All around us, life quickens and stirs. I am conscious of how all creatures, in their wonderfully complex relationships, are entangled in the inanimate world. Gravity gives them direction, the ground supports them, the atmosphere nourishes them, and the Sun warms them. The Earth and the sky and space, out to the far edge of the Solar System, and beyond that to the limits of the universe, are all part of a single unity. The dim intuition of that momentous realization provides the impetus for science, art, and spirituality.

As we approach the parking lot at the end of the road I remember the killdeer. One of these days, very soon now, the chicks will hatch, and we will see them darting around like partially toasted marshmallows on their toothpick legs, calling to their mother with feeble, shrill cheeps. When we get to the barrier, I put the leash back on Ted. I wouldn't want him to get lucky.

THE THEORY
OF THE UNIVERSE?

MICHIO KAKU

When I was a child of eight, I heard a story that will stay with me for the rest of my life. I remember my schoolteachers telling us about a great scientist who had just died. They talked about him with great reverence, calling him one of the greatest scientists in all history. They said that very few people could understand his ideas, but that his discoveries had changed the entire world and everything around us.

But what most intrigued me about this man was that he had died before he could complete his greatest discovery. They said he had spent years on this theory, but he died with unfinished papers still sitting on his desk.

I was fascinated by the story. To a child, this was a great mystery. What was his unfinished work? What problem could possibly be so difficult and so important that such a great scientist would dedicate years of his life to its pursuit?

Curious, I decided to learn all I could about Albert Einstein and his unfinished theory. Some of the happiest moments of my childhood were spent quietly reading every book I could find about this great man and his ideas. When I exhausted the books in our local library, I began to scour libraries and bookstores across the city and state, eagerly searching for more clues. I soon learned that this story was far more exciting than any murder mystery and more important than anything I could ever imagine. I decided that I would try to get to the root of this mystery, even if I had to become a theoretical physicist to do it.

Gradually, I began to appreciate the magnitude of his unfinished quest. I learned that Einstein had three great theories. The first two, the special and the general theories of relativity, led to the development of the atomic bomb and to our present-day conceptions of black holes and the Big Bang. These two theories by themselves earned him his reputation as the greatest scientist since Isaac Newton.

However, Einstein was not satisfied. The third, which he called the *unified field theory,* was to have been his crowning achievement. It was to be the theory of the universe, the Holy Grail of physics that would finally unify all physical laws into one simple framework. It was to have been the ultimate goal of all physics, the theory to end all theories.

Sadly, it consumed Einstein for the last thirty years of his life; he spent many lonely years in a frustrating pursuit of the greatest theory of all time. But he wasn't alone; I learned that some of the greatest minds of the twentieth century, such as Werner Heisenberg and Wolfgang Pauli, also struggled with this problem and ultimately gave up.

Given the fruitless search that has stumped these and other Nobel Prize winners for half a century, most physicists agree that the Theory of Everything must be a radical departure from everything that has been tried before. For example, when Niels Bohr, founder of modern atomic theory, once listened to Pauli's explanation of his own version of the unified field theory, Bohr finally stood up and said, "We are all agreed that your theory is absolutely crazy. But what divides us is whether your theory is crazy enough."

Today, however, after decades of false starts and frustrating dead ends, many of the world's leading physicists think that they have finally found the theory "crazy enough" to be the unified field theory. Scores of physicists in the world's major research laboratories now believe we have at last found the Theory of Everything.

The theory that has generated so much excitement is called the *superstring theory.* Nearly every science publication in the world has featured major stories on the superstring theory, interviewing some of its pioneers, such as John Schwarz, Michael Green, and Yoichiro Nambu. (*Discover* magazine even featured it twice on its cover.) My book *Beyond Einstein: The Cosmic Search for the Theory of the Universe* was the first attempt to explain this fabulous theory to the lay audience.

Naturally, any theory that claims to have solved the most intimate secrets of the universe will be the center of intense controversy. Even Nobel Prize winners have engaged in heated discussions about the validity of the superstring theory. In fact, over this theory we are witnessing the liveliest debate in theoretical physics in decades.

To understand the power of the superstring theory and why it is heralded as the theory of the universe (and to understand the delicious controversy that it has stirred up), it is necessary to understand that there are four forces that control everything in the known universe, and that the superstring theory gives us the first (and only) description that can unite all four forces in a single framework.

THE FOUR FUNDAMENTAL FORCES

Over two thousand years ago, the ancient Greeks thought that all matter in the universe could be reduced to four elements: air, water, earth, and fire. Today, after centuries of research, we know that these substances are actually composites; they in turn are made of smaller atoms and subatomic particles held together by just four and only four fundamental forces.

Gravity is the force that keeps our feet anchored to the spinning earth and binds the solar system and the galaxies together. If the force of gravity could somehow be turned off, we would be immediately flung into outer space at approximately a thousand miles per hour. Furthermore, if gravity did not hold the Sun together, it would explode in a catastrophic burst of energy. Without gravity, the Earth and the planets would spin out into freezing deep space and the galaxies would fly apart.

Electromagnetism is the force that lights up our cities and energizes our household appliances. The electronic revolution, which has given us the light bulb, TV, the telephone, the computer, radio, radar, the microwave, and the dishwasher, is a byproduct of the electromagnetic force. Without this force, our civilization would be wrenched several hundred years into the past, into a primitive world lit by candlelight and camp fires.

The strong nuclear force is the force that powers the Sun. Without the nuclear force, the stars would flicker out and the heavens would go dark. Without the Sun, all life on Earth would perish as the oceans turned to solid ice. The nuclear force not only makes life on Earth possible, it is also the devastating force unleashed by a hydrogen bomb, which can be compared to a piece of the Sun brought down to Earth.

The weak nuclear force is the force responsible for radioactive decay. The weak force is harnessed in modern hospitals in the form of radioactive tracers used in nuclear medicine. For example, dramatic color pictures of the living brain as it thinks and experiences emotions are made possible by the decay of radioactive sugar in the brain.

It is no exaggeration to say that the mastery of each of these four fundamental forces has changed every aspect of human civilization. For example, when Newton tried to solve his theory of gravitation, he was forced to develop a new mathematics and formulate his celebrated laws of motion. These laws of mechanics in turn helped to usher in the Industrial Revolution.

Furthermore, the mastery of the electromagnetic force by mathematical physicist James Maxwell in the 1860s has revolutionized our way of life. Whenever there is a power blackout, we are forced to live much like our forebears in the last century. Today, over half of the world's industrial wealth is connected, in some way or other, to the electromagnetic force, without which modern civilization is unthinkable.

Similarly, when the nuclear force was unleashed with the atomic bomb, human history for the first time faced a new and frightening set of possibilities, including the total annihilation of all life on Earth. With the nuclear force, we could finally understand the enormous engine that lies within the Sun and the stars, but we could also glimpse for the first time the end of humanity itself.

Thus, whenever scientists unravel the secrets of one of the four fundamental forces, they irrevocably alter the course of modern civilization. Some of the greatest breakthroughs in the history of the sciences can be traced back to the gradual understanding of these forces.

Given their importance the next question is, Can these four fundamental forces be united into one super force? Are they but diverse manifestations of a deeper reality?

TWO GREAT THEORIES

At present there are two physical frameworks that have partially explained the mysterious features of these four fundamental forces. Remarkably, these two formalisms, the *quantum theory* and *general relativity,* allow us to explain the *sum total of all physical knowledge* at the fundamental level. Without exception.

All the laws of physics and chemistry, which can fill entire libraries with technical journals and books, can in principle be derived from these two fundamental theories—making these the most successful physical theories of all time, withstanding the test of thousands of experiments and challenges.

Ironically, these two fundamental frameworks are diametrically opposed to each other. The quantum theory, for example, is the theory of the microcosm, with unparalleled success at describing the subatomic world. The theory of relativity, by contrast, is a theory of the macrocosmic world, the world of galaxies, superclusters, black holes, and Creation itself.

The quantum theory explains three of the four forces (the weak and strong nuclear forces, and the electromagnetic force) by postulating the exchange of tiny packets of energy, called *quanta*. When a flashlight is turned on, for example, it emits trillions upon trillions of photons, or quanta, of light. Lasers, radar waves, and microwaves all can be described by postulating that they are caused by the movement of these tiny quanta of energy. Likewise, the weak force is governed by the exchange of subatomic particles called *W-bosons*. The strong nuclear force, in turn, binds protons together by the exchange of *gluons*.

However, the quantum theory stands in sharp contrast to Einstein's general theory of relativity, which postulates an entirely different physical picture to explain the force of gravity.

Imagine, for the moment, dropping a heavy shotput on a large bedspread. The shotput will, of course, sink deeply into the bedspread. Now imagine shooting a small marble across the bed. Since the bed is warped, the marble will execute a curved path. However, for a person viewing the marble from a great distance, it will appear that the shotput is exerting an invisible "force" on the marble, forcing it to move in a curved path. In other words, we can now replace the clumsy concept of a "force" with the more elegant concept of a bending of space itself. We now have an entirely new definition of this "force." It is nothing but the byproduct of the warping of space.

In the same way that a marble moves on a curved bedspread, the Earth moves around the Sun in a curved path, because space-time itself is curved. In this new picture, gravity is not a "force" but a byproduct of the warping of space-time. In some sense, gravity does not exist; what moves the planets and stars is the distortion of space and time.

However, the problem that has stubbornly resisted solution for fifty years is that these two frameworks do not resemble each other in any way. The quantum theory reduces "forces" to the exchange of discrete packets of energy, or quanta, while Einstein's theory of gravity, by contrast, explains the cosmic forces holding the galaxies together by postulating the smooth deformation of the fabric of space-time. This is the root of the problem, that the quantum theory and general relativity have two different physical pictures (packets of energy versus smooth space-time continua) and different mathematics to describe them. This sad state of affairs can be compared to Mother Nature having two hands, neither of which communicates with the other.

All attempts by the greatest minds of the twentieth century at merging the quantum theory with the theory of gravity have failed. Unquestionably, the greatest problem facing physicists today is the unification of these two physical frameworks into one theory.

SUPERSTRINGS

Today, however, many physicists think that we have finally solved this long-standing problem. A new theory, which is certainly "crazy enough" to be correct, has astounded the world's physics community. But it has also raised a storm of controversy, with Nobel Prize winners adamantly taking opposite sides of the issue.

This is the superstring theory, which postulates that all matter and energy can be reduced to tiny strings of energy vibrating in a ten-dimensional universe.

Edward Witten, of the Institute for Advanced Study at Princeton, who some claim is the successor to Einstein, has said that superstring theory will dominate the world of physics for the next fifty years, in the same way that the quantum theory has dominated physics for the last half century.

As Einstein once said, all great physical theories can be represented by simple pictures. Similarly, superstring theory can be explained visually. Imagine a violin string, for example. The note A is no more fundamental than the note B. What is fundamental is the violin string itself. By studying vibrations or harmonies on a violin string, one can calculate the infinite number of possible frequencies that can exist.

Similarly, the superstring can also vibrate in different frequencies. Each frequency, in turn, corresponds to a subatomic particle, or a quantum. This explains why there appears to be an infinite number of particles. According to this theory, our bodies, which are made of subatomic particles, can be described by the resonances of trillions upon trillions of tiny strings.

In summary, the "notes" of the superstring are the subatomic particles, the "harmonies" of the superstring are the laws of physics, and the universe can be compared to a "symphony" of vibrating superstrings.

As the string vibrates, however, it causes the surrounding space-time continuum to warp around it. Miraculously enough, a detailed calculation shows that the superstring forces the space-time continuum to be distorted exactly as Einstein originally predicted. Thus, we now have a harmonious description that merges the theory of quanta with the theory of space-time continua.

TEN-DIMENSIONAL HYPERSPACE

The superstring theory represents perhaps the most radical departure from ordinary physics in decades. But its most controversial prediction is that the universe originally began in ten dimensions. To its supporters, the prediction of a ten-dimensional universe has been a conceptual tour de force introducing a startling, breathtaking mathematics into the world of physics. To its critics, however, the introduction of ten-dimensional hyperspace borders on science fiction.

To understand these higher dimensions, we must remember that it takes three numbers to locate every point in the universe, from the tip of your nose to the ends of the universe.

For example, if you want to meet some friends for lunch in Manhattan, you say that you will meet them at the building at the corner of Forty-second and Fifth Avenue, on the thirty-seventh floor. It takes two numbers to locate your position on a map, and one number to specify the distance above the map.

However, the existence of the fourth spatial dimension has been a lively area of debate since the time of the Greeks. Ptolemy, in fact, even gave a "proof" that more than three dimensions cannot exist. Ptolemy reasoned that only three straight lines that are mutually perpendicular can be drawn (for example, the three perpendicular lines making up a corner of a room). Since a fourth straight line cannot be drawn perpendicular to each of the other three axes—ergo!—the fourth dimension cannot exist.

What Ptolemy actually proved was that it is impossible for us to *visualize* the fourth dimension. Although computers routinely manipulate equations in *n*-dimensional space, humans are incapable of visualizing more than three dimensions.

The reason for this unfortunate accident has to do with biology rather than physics. Human evolution put a premium on being able to visualize objects moving in three dimensions, such as lunging saber-tooth tigers and charging mammoths.

Since tigers do not attack us in the fourth dimension, there was no evolutionary correction pressure to develop a brain with the ability to visualize four dimensions.

From a mathematical point of view, however, adding higher dimensions has a distinct advantage: It allows us to describe more forces. There is more "room" in higher dimensions to insert the electromagnetic force into the gravitational force. (In this picture, light becomes a vibration in the fourth dimension.) In other words, adding more dimensions to a theory always allows us to unify more laws of physics.

A simple analogy may help. The ancients were once puzzled by the weather. Why does it get colder as we go north? Why do the winds blow to the west? What is the origin of the seasons? To the ancients, these were mysteries that could not be solved.

The key to these puzzles, of course, is to leap into the third dimension, to go *up* into outer space, to see that the Earth is actually a sphere rotating around a tilted axis. In one stroke, these mysteries of the weather—the seasons, the winds, the temperature patterns, etc.—become transparent.

Likewise, the superstring is able to accommodate a large number of forces because it has more "room" in its equations to do so.

WHAT HAPPENED BEFORE THE BIG BANG?

One of the nagging problems of Einstein's old theory of gravity was that it did not explain the origin of the Big Bang.

The ten-dimensional superstring theory, however, gives us a compelling explanation according to which the universe originally started as a perfect ten-dimensional universe with nothing in it.

However, this ten-dimensional universe was not stable. The original ten-dimensional space-time finally "cracked" into two pieces, four- and six-dimensional universes. The six-dimensional universe collapsed into a tiny ball, while the remaining four-dimensional universe inflated at an enormous rate.

The four-dimensional universe (ours) expanded rapidly, eventually creating the Big Bang, while the six-dimensional universe wrapped itself into a ball and collapsed down to infinitesimal size.

The Big Bang is now viewed as a rather minor aftershock of a more cataclysmic collapse: the breaking of a ten-dimensional universe into four- and six-dimensional universes.

In principle, it also explains why we cannot measure the six-dimensional universe: it has shrunk down to a size smaller than an atom.

RE-CREATING CREATION

Although the superstring theory has been called the most sensational discovery in theoretical physics in the past decades, its critics have focused on its weakest point, that it is almost impossible to test. The energy at which the four fundamental forces merge into a single unified force is the fabulous *Planck energy*, which is a billion billion times greater than the energy found in a proton.

Even if all the nations of the Earth were to band together and single-mindedly build the biggest atom smasher in all history, it would still not be enough to test this theory.

Because of this, some physicists have scoffed at the idea that superstring theory can be considered a legitimate theory. Nobel laureate Sheldon Glashow, for example, has compared the superstring theory to the former President Reagan's Star Wars program because it is untestable and drains the best scientific talent.

The reason the theory cannot be tested is rather simple. The Theory of Everything is necessarily a theory of Creation. It must explain everything, from the origin of the Big Bang down to that of the lilies of the field. To test this theory on Earth, therefore, means to re-create Creation on Earth, which is impossible with present-day technology.

THE SSC: BIGGEST EXPERIMENT OF ALL TIME

These questions about unifying the fundamental forces may not be academic if the largest scientific machine ever, the SSC, is built to test some of our ideas about the instant of Creation. (Although the SSC was originally approved by the Reagan administration, because of its enormous cost, the project is still touch-and-go, depending every year on Congressional funding.)

The SSC is projected to accelerate protons to a staggering energy of tens of trillions of electron volts. When these subatomic particles slam into each other at these fantastic energies, they will generate temperatures that have not been reached since the instant of Creation (although not hot enough to test fully the superstring theory). That is why the supercollider is sometimes called a "window on Creation."

The SSC is projected to cost over eight billion dollars (a large amount of money compared to the government's science budget, but insignificant relative to that of the Pentagon). By every measure, it will be a colossal machine. It will consist of a ring of powerful magnets stretched out in a tube over fifty miles in diameter. In fact, one could easily fit the Washington Beltway, which surrounds Washington, D.C., inside the SSC.

At present, the SSC is scheduled to be finished near the turn of the century in Texas, near the city of Dallas. When completed, it will employ thousands of physicists and engineers and cost millions of dollars to operate.

At the very least, physicists hope that the SSC will find some exotic subatomic particles, such as the Higgs boson and the top quark, in order to complete our present-day understanding of the quantum theory. However, there is also the small chance that physicists might discover "supersymmetric" particles, which are predicted by the superstring theory. In other words, although the superstring theory cannot be tested directly by the SSC, one hopes to find particles (vibrations) predicted by superstring theory among the debris created by smashing protons together.

THE PARABLE OF THE GEMSTONE

To understand the intense controversy surrounding superstring theory, think of the following parable.

Imagine that at the beginning of time there was a beautiful, glittering gemstone. Its perfect symmetries were a sight to behold. However, it possessed a tiny flaw and became unstable, eventually exploding into thousands of pieces. Imagine that the fragments of the gemstone rained down on a flat, two-dimensional world called Flatland, where there lived a mythical race of beings called Flatlanders.

These Flatlanders were intrigued by the beauty of the fragments, which could be found scattered all over Flatland. The scientists of Flatland postulated that these fragments must have come from a crystal of unimaginable beauty that shattered in a titanic Big Bang. They then decided to embark upon a noble quest to reassemble all the pieces of the gemstone.

After two thousand years of labor by the finest minds of Flatland, they were finally able to fit many, but certainly not all, of the fragments together in two chunks. The first chunk was called the *quantum,* and the second chunk was called *relativity.*

Although the Flatlanders were rightfully proud of their progress, they were dismayed to find that these two chunks did not fit together. For half a century, the Flatlanders maneuvered the chunks in all possible ways and still could not make them fit.

Finally, some of the younger, more rebellious scientists suggested a heretical solution: perhaps these two chunks could fit together if they were moved in a *third dimension.*

This immediately set off the greatest scientific controversy in years. The older scientists scoffed at this idea, because they didn't believe in an unseen third dimension. "What you can't measure doesn't exist," they declared.

Furthermore, even if the third dimension existed, one could calculate that the energy necessary to move the pieces *up* off Flatland would exceed all the energy available in Flatland. Thus it was an untestable theory.

However, the younger scientists were undaunted. Using pure mathematics, they could show that these two chunks would likely fit together if they were rotated and moved in the third dimension. The younger scientists claimed that the problem was therefore theoretical rather than experimental. If one could completely solve the equations of the third dimension, one could, in principle, fit the two chunks perfectly together and resolve the problem once and for all.

WE ARE NOT SMART ENOUGH

That is also the conclusion of today's superstring enthusiasts: the fundamental problem is theoretical, not practical. The true problem is in solving the theory completely and then comparing it with present-day experimental data, not in building gigantic atom smashers.

Edward Witten, impressed by the vast new areas of mathematics opened up by the superstring theory, has said that the superstring theory represents "twenty-first-century physics that fell accidentally into the twentieth century."

The superstring theory may very well be twenty-first-century physics, but twenty-first-century mathematics has not yet been discovered.

This situation is not entirely new to the history of physics. When Newton first discovered the universal law of gravitation at the age of twenty-three, he was unable to solve his equation because the mathematics of the seventeenth century was too primitive. He then labored over the next twenty years to develop a new mathematical formalism (calculus) that was powerful enough to solve his universal law of gravitation.

Similarly, the fundamental problem facing the superstring theory is theoretical. If we could only sharpen our analytical skills and develop more powerful mathematical tools, perhaps we could solve the superstring theory and end the controversy.

Ironically, the superstring equations stand before us in perfectly well defined form, yet we are too primitive to understand why they work so well, and we are too dim-witted to solve them. The search for the theory of the universe is perhaps finally entering its last phase, awaiting the birth of a new mathematics powerful enough to solve it.

Imagine a child gazing at a TV set. The images and stories conveyed through the screen are easily understood by the child, yet the electronic wizardry inside the TV set is beyond the child's ken. Likewise, we physicists gaze in wonder at the mathematical sophistication and elegance of the superstring equation and are awed by its power, yet we do not understand why it works.

Perhaps some readers will be inspired by this story to read every book in their libraries about the superstring theory. Perhaps some young reader will be the one to complete this quest for the theory of the universe, begun so many years ago by Einstein.

The Law of Gravity, the Law of Levity, and Murphy's Law

Judith Stone

Perhaps the most potent illustration of Murphy's Law—"If anything can go wrong, it will"—is Murphy's Law. Because that's not what Murphy said.

His actual words were, "If there is more than one way to do something and one of those ways won't work, somebody is going to come along and try it."

As predicted by the popularized version of his eponymous postulate (snappier, perhaps, than the *Ur*-utterance, but less rich in both nitty and gritty), my search for Edward Aloysius Murphy, Jr., hit a major snag: by the time I discovered that he was no mere myth and had, in fact, a telephone number, Murphy had exited this flawed world whose essence he'd so accurately captured in his enthusiastically, though erroneously, adopted aphorism. But his gracious widow, Mrs. Effie Murphy, gave me the official history.

After serving in World War II, Captain Ed Murphy, West Point graduate, bomber pilot and engineer, became chief of the structures lab at Wright Air Development Center in Ohio, testing aircraft and developing crash-proof seats. In 1946, he joined Project MX981 with Major John Paul Stapp of Edwards Air Force Base in California's Mojave Desert. They were both working on safety equipment that could help pilots survive crashes—Murphy in the lab, Stapp riding a rocket-powered sled down a steel track to simulate the rapid acceleration, then deceleration, of a crash.

"Stapp's goal was to achieve a land speed of 640 miles per hour, then stop within 1.3 seconds and survive the stop," wrote Murphy in an informational handout he prepared for the curious later in his life. In 1949, just before a key test, Stapp couldn't get a reading from his accelerometer, an electronic device that measured the tensile force a shoulder harness and safety belt would have to sustain in order to save a pilot in a simulated air crash. "He called the designer of the accelerometer, Murphy, and invited him to California to solve the accelerometer dilemma." Murphy discovered that a technician had installed a vital gauge backward.

"Later Stapp asked Murphy to explain what caused the foul-up. Murphy's now famous reply was 'If there is more than one way to do something and one of those ways won't work, somebody is going to come along and try it.' Meaning: make sure you consider all the alternatives, especially the catastrophic ones."

At a press conference not long after the incident, Stapp attributed the project's fine safety record to a strong belief in Murphy's Law. A reporter asked Stapp about the law, which he rendered as "If anything can go wrong, it will." Aerospace industry advertising picked up the mangled trope, and it captured the global imagination.

From the first, pop-Murphy drew eager disciples who formulated corollaries to the master's primal (misdisseminated) pronouncement. Among the embellishments: Murphy's Second Law: "Nothing is as easy as it looks"; Murphy's Third Law: "Everything takes longer than you think it will"; Mrs. Murphy's Law: "Anything that can go wrong will go wrong while Murphy's out of town"; Murphy's Law of Thermodynamics: "Things get worse under pressure."

Murphy's Law became, in fact, the subject of several best-selling books and a popular motif for posters, mugs, and T-shirts, although Murphy never made a cent from them. Arthur Bloch, the author of *Murphy's Law,* mentioned Ed Murphy: "But what was popularized as Murphy's Law, the version I used, he didn't even say—and it was just the jumping-off point for a collection of five hundred to a thousand other laws." (Like Osborn's Law: "Variables won't, constants aren't." And Cole's Law: "Thinly sliced cabbage.") "Ed thought about suing," says Mrs. Murphy, "but he was too easygoing a fellow, and he decided life was too short for such trouble."

Twenty years ago, Paul Dickson of Garrett Park, Maryland, founded the Murphy Center for the Codification of Human and Organizational Law, actually a shoe box full of odd Murphy's-Law-like rules he'd collected. He's since published three books' worth—the latest is *The New Official Rules*—and still gets hundreds of letters a week with offerings like Murphy's Flaw: "If anything can't go wrong [i.e., Murphy's Law] it will"; Smith's Fourth Law of Inertia: "A body at rest tends to watch television"; The First Law of Sociogenetics: "If your parents didn't have children, chances are you won't either"; and my own modest effort, Stone's Constant: "The one day you don't wear makeup is the day you run into your ex-husband."

Certainly the notion that the cosmic deck is stacked against us is not a new one. The spirit of Murphy-as-popularized is at least four thousand years old: an ancient Egyptian poem "The Man Who Was Tired of Life," thought by scholars to have been written in 1990 B.C., reads, in part, "The wrong that roams the earth, there is no end to it." Julius Caesar is reported to have said, *"Quod malum posset futurum,"* which translates roughly as "What is bad will come to pass," pure pop-Murphy.

The folk wisdom of many cultures shows a distinctly Murphyan strain. "The hidden stone finds the plow," sigh the Estonians. "The spot always falls on the best cloth," the Spanish say. "Even an unloaded rifle can fire once in ten years; and once in one hundred years, even a rake can produce a shot," the Russians insist (but not very convincingly). Apparently there's a British version of Murphy, one Sod, given to formulating similarly grim theorems. (Sod's Law: "The degree of failure is in direct proportion to the effort expended and the need for success.") An expert on Sod, Richard Boston, of London, has traced such sentiments to a proverb that first appeared in print in 1871: "The bread never falls but on its buttered side." (According to a report in the *Listener* of Great Britain, researchers found that toast falls buttered-side down six times out of ten, because of the weight of the butter. I tried to replicate the study, but failed to figure in the drag coefficient of pumpernickel. Murphyed again.)

Murphy's Law makes complete sense if you agree that life is a comedy to those who think and a tragedy to those who try to assemble their own furniture. It's certainly an easier law to remember than Newton's, Boyle's, or Gresham's. (Which one goes, "Bad money drives out gas"?) Murphy's hits the zeitgeist on the head (to mix, blend, and liquefy a metaphor); it is a maxim made to order for the waning days of the twentieth century, an epoch that has almost put satire out of business.

Human factors engineer Clifford Wong, of the California aerospace firm McDonnell Douglas, is an expert on human-machine interaction. He finds pop-Murphy, though meant joshingly, to be helpful. "I think it keeps designers and engineers on our toes, reminding us to consider potential disasters and avoid them." Sociologist Laurence Peter believes it provides a healthy laugh. "When things go wrong, it gives you a humorous escape." Peter's famed principle, published in 1969, has, like Murphy's Law, been widely accepted: "In a hierarchy," he posited, "every employee tends to rise to his level of incompetence." (The Peter Principle is not to be confused with Parkinson's Law, the 1955 dictum of historian C. Northcote Parkinson: "Work expands so as to fill the time available for its completion.") Stone's Corollary to the Peter Principle is the Paul and Mary Principle: "On a camping trip, the jerk who insists that everyone sing 'Michael Row the Boat Ashore' doesn't know any of the words except 'Alleluia.' " (Stone's First Law of Levity: "Things that make me laugh so hard that beer comes out my nose leave throngs of others unmoved.")

Sure, invoking Murphy may be an amusing distraction while we're waiting to be done in by the Greenhouse Effect (the phenomenon of global warming), the White House Effect (the phenomenon of ignoring global warming), or some as yet unnicknamed catastrophe. Gallows humor can be a release. But it can also be a way of avoiding responsibility.

Says Murphy's oldest son, Edward, a lawyer in Sausalito, California,

"My father formulated not a law of probability but a law of failure—of how things could fail if people didn't take care. For his aerospace work —he went on to be involved in the *Apollo* and *Gemini* programs as an aeronautical design engineer—my father performed test procedures over and over, sometimes for years, to find the smallest flaw. He and his colleagues were trying to reduce the probability of failure to a low level, and they did.

"I know one of the things my father *didn't* consider an illustration of Murphy's Law, as popularized: the explosion of the space shuttle *Challenger*. And I think my father would say, 'If you maintain a proper hold on toast, it doesn't fall to the floor.' "

Murphy, who died in 1990 at the age of seventy-two, took the misquote good-naturedly, his wife says, but worried that it would breed irresponsibility. "His original statement wasn't fatalistic. He stressed that he meant, 'If there's any way for a person to do something wrong, a person will find it.' It isn't just fate; there has to be a human goofus involved somewhere. He didn't want the famous version of his law to encourage carelessness or allow people to rationalize their failures."

That fear is echoed by Yale University sociologist Charles Perrow, author of *Normal Accidents,* a study of risk assessment. "I think Murphy's Law, as popularized, is psychologically dangerous," he says. "It keeps people from working harder to avoid problems, to make changes, and to guard against carelessness. If you amend the law slightly to 'If a system is complex enough, there are bound to be failures; therefore, don't build such complex systems,' we'd be better off. In any case, Murphy's interpreters are wrong, because most things go right most of the time. What an amazing achievement that some eighty million Americans get to work every morning! Except perhaps on Monday."

It's human nature to notice when things go wrong and not when they go right, the same quirk that makes us forget a compliment in a matter of minutes but nurse an insult for generations. How many of us shake a fist at the heavens, demanding an explanation for our good fortune? This tendency may be evolutionary; perhaps protohominid optimists were stoned by the rest of the tribe for their nauseating Pollyannaism. In any case, Murphy's Law and its history resonate with Oppenheimer's Observation, made by J. Robert Oppenheimer in the *Bulletin of the Atomic Scientists* in 1951: "The optimist thinks this is the best of all possible worlds, and the pessimist knows it is."

TIME REVERSAL

A. ZEE

Recently, I went to Michigan to speak at a conference devoted to time reversal.

Time reversal? Yes, time reversal. "Wow!" you say. "Reversing the flow of time?"

Perhaps no other term in modern physics inspires as much fascination and awe in the general public as time reversal. Consider the enormous popularity of such recent movies as *Back to the Future I, II,* and *III, Terminator 1,* and *Peggy Sue Got Married.* But those people curious enough to trudge over to a library and look up time reversal in an advanced physics text—for instance, chapter four in the classic by J. J. Sakurai, which I studied as a student, bears the dramatic title "Time Reversal"— are bound to be disappointed. Physicists do not know how to arrange for you to go back to your high-school days and see teen romance with the wisdom of an adult, as Kathleen Turner did. Nor can they send you or Michael J. Fox back in time to help your father beat out his competitor for the affection of your mother. Sorry to disappoint you, but we physicists have not the foggiest idea how to reverse the flow of time.

Physicists are focusing on a much more modest question, a question you have to ask before you can wonder whether or not you can go back in time. They are trying their damnedest to find out whether the fundamental laws of physics know about the "arrow of time."

We are all aware of the psychological arrow of time, a subjective feeling that time flows mercilessly from the past to the future. Physicists also speak of a thermodynamic arrow of time along which physical systems invariably become ever more disordered. Perhaps the physiological arrow of time, measured by the aging of our bodies, is just a manifestation of this thermodynamic arrow of time. The expansion of the universe provides yet another arrow of time. The big question is whether, and how, these arrows are related to each other. In particular, we would all like to know how our psychological arrows of time come about, and then, of course, whether we can reverse these arrows.

Well, dear reader, an incredible amount of quasiphilosophical speculation and mumbo jumbo has been uttered about these arrows of time, the kind of mumbo jumbo most physicists are just not smart enough to understand. In the time-honored tradition of their profession, physicists tend to want to answer one question at a time, starting with the simplest.

So, let's start with a simple one. Do the laws of physics contain in them an arrow of time?

Physicists have the curious habit, when a question is posed to them, of demanding that the questioner outline a procedure—a series of operations, if you will—so that by following that procedure, they can actually answer the question one way or another. I strongly recommend this attitude to the world at large. This operational approach is guaranteed to save plenty of time and energy, not to mention hot air. Okay, you want to know how many angels can dance on the head of a pin. Just tell me a procedure whereby I can determine the answer, even if only in principle and not in actual practice. First, find some angels . . . You get the point. Unfortunately for me, I got a terrible grade in the philosophy course I took in college because I kept pestering the professor. "Give me an operational procedure to follow so I can determine whether what you just told me is true or not."

I will now give you the operational formulation of the question I mentioned above. Make a movie of a physical process. Run the movie backward. Are the events portrayed in the backward-running movie allowed by the laws of physics? In other words, can a bunch of physicists, simply by applying their professional expertise, determine whether a movie is being played forward or backward? If not, then they say that the laws governing that process are invariant under time reversal: those laws don't know beans about an arrow of time.

"This is ludicrous," you say, "now that I understand what you pinhead physicists are talking about. Of course, the laws of physics are not invariant under time reversal."

You invite a bunch of said pinhead physicists over to your place, pick out a videotape from your collection of vintage baseball clips, and stick it into your VCR. Ty Cobb ambles up to the plate, swings—yes!—the bat makes solid contact with the ball. *Bam, wham.* Home run! You press a button and the videotape reverses itself. A ball comes flying in from outside the stadium. It moves faster and faster—*whack!*—it smacks the bat and Ty Cobb's arm backward. Of course, you can tell whether the movie is played forward or backward; you smirk in triumph.

"Not so fast," the physicists reply as a chorus. Sure, it is improbable that a ball could come flying in from the parking lot and hit the bat, but it is not impossible. It is not forbidden by the laws of physics. A ball moving at just the right speed could push the bat backward.

But no, now it is your turn to object: "When I play the videotape backward, the ball flies in faster and faster. Surely that is forbidden by the known laws of physics."

"Yeah," yell the physicists, "that's because you forgot the motion of the air molecules. If your movie were sharp enough to show all relevant details, you would see that after Ty Cobb hits that ball, the ball keeps colliding with air molecules and thus slows down as it flies out of the park. When you play the movie backward, we would have seen—if your movie were sharp enough—zillions of air molecules conspiring to collide with the ball in precisely such a way as to push the ball faster and faster toward the bat. Perhaps just to fool us, you have somehow diabolically arranged for the air molecules to do precisely that."

The physicists, most of them being university professors, probably cannot resist giving you a didactic lecture about friction at this point. In the everyday world of macroscopic objects, we see friction working to slow things down, and this seemingly provides an arrow of time. But that is of course due to processes (such as collisions of the ball with air molecules in our example) that we have neglected to keep track of.

My somewhat silly story is meant to emphasize that to study time reversal, physicists simply examine one physical process after another and see whether the time-reversed process is allowed by the laws of physics. They soon realized, obviously enough, that it is not necessary to look at such a complicated process as a home run. For the purpose of studying the time-reversal invariance of the laws of physics, it is enough to reduce complicated processes to ever-simpler processes that, when put together, make up whichever complicated processes happen to interest us. And so from studying the collision between ball and bat, physicists move on to studying the collisions between molecules, between atoms, and between subnuclear particles.

For example, back in 1951, some physicists at the University of California at Berkeley crashed two protons together and observed a pion and a deuteron coming out of the collision. (For the purpose of our discussion here, you don't have to know what these particles are.) The time-reversed movie shows a pion and a deuteron colliding and producing two protons. So physicists at Columbia University and the University of Rochester did precisely that: they crashed a pion into a deuteron and observed two protons coming out. What they saw looked exactly as if they were watching the movie of their California colleagues' experiment played backward. The laws governing that particular process appeared to be perfectly time-reversal invariant.

Over the last few decades, physicists have watched a lot of "movies" played forward and backward, and they have never seen hide or hair of an arrow of time anywhere in the fundamental laws of physics. Very strange, given that we see arrows of time all around us.

Of course, physicists are never so dumb as to say that they don't see something. They say, "We don't see this or that at such and such degree of accuracy as limited by our apparatus, by how much money the people in Washington would give us," and so on. Over the years, some extraordinarily persistent and hard-working experimenters have stretched their ingenuity to the limit to improve the accuracy with which they do not see an arrow of time in the laws of physics.

Physicists soon realized that crashing particles together was not the best way to study time reversal. In our example, the East Coast experimenters had to make sure that the pion and deuteron came together with precisely the same energy and momentum as those with which they were observed to come out on the West Coast. In the real world, you just can't arrange that to a high degree of accuracy.

In 1956, physicists found a better way to go. To explain this, I have to ask you to remember how much fun it was to spin a top when you were a child. (In this day and age of video games, alas, top spinning is going out of our collective cultural experience. Sad, sad, but that's the subject of another article.)

Every child knows that as a top spins, the axis around which it spins moves around in a circle: it *precesses*. Let us film the spinning top. Now run the movie backward. We see the top precessing in the opposite direction, but the top is spinning in the opposite direction, too. If you are an observant child, you notice that if you spin the top one way, it precesses one way, and if you spin the top the other way, it precesses the other way. In other words, you determine that the law governing the motion of the top is invariant under time reversal. (Well, I for one certainly was not *that* observant.)

As a matter of fact, the electron also spins. Because there is no friction in the microscopic world, an electron just keeps on spinning at the same rate forever. The spinning electron acts like a little magnet, and in a magnetic field it precesses just like the spinning top precesses in the earth's gravitational field. This precession of the electron in a magnetic field is observed routinely in the laboratory—no big deal. The direction in which the electron precesses depends on the direction of the magnetic field. If we reverse the magnetic field, the electron precesses the other way.

Does the precessing electron know about the arrow of time? Those physicists, they immediately made a movie and ran it backward—they are paid to do things like that. If you took physics in high school, you might remember that a magnetic field is produced only by moving electric charges. Charges that are sitting there produce not a magnetic field but only an electric field. (In a bar magnet, the charged particles in the iron are constantly moving around in an organized fashion. In the modern laboratory, magnetic fields are usually produced by electric currents, namely electric charges flowing in a wire.)

Okay, after this little bit of high-school physics, we can now turn on the projector, put in a movie of an electron precessing, and play it backward. The electron precesses the other way. We get excited! Can this be an arrow of time? By watching the direction in which the electron precesses, can we tell which way the movie is being played? But wait, the magnetic field also reverses direction, because the moving charges that produce the magnetic field move the other way in the backward-running movie. We can't tell whether the movie is played forward or backward. No arrow of time anywhere in sight.

Flash! The lightbulb goes on in your head. How about an electron precessing in an electric field? If you make a movie of that and play it backward, the electric field doesn't reverse direction. Why? Because it is produced by charges just sitting there. But the electron precesses the other way. The forward-running movie and the backward-running movie look different: the electric fields point in the same direction in both, but the electrons precess in different directions. You would have found an arrow of time if you saw an electron precessing in an electric field.

You rush to your lab, elbowing your colleagues out of the way. You set up an electric field and stick an electron in it. The instant you see it precessing, you can buy that ticket to Stockholm! Fame and glory for him or her who finds an arrow of time in the fundamental laws of physics!

I just gave you a rather long-winded explanation, with all that high-school physics thrown in—I am a physics professor, after all. But I want to emphasize that the basic idea couldn't be simpler. Under time reversal, magnetic fields reverse directions, but electric fields do not. If the laws of physics are invariant under time reversal, then the electron cannot possibly precess in an electric field as well as in a magnetic field.

A bit of terminology here: a particle that precesses in an electric field is said to have an *electric dipole moment*. So far I have focused on the electron, but nothing in the discussion depends specifically on the electron. If any particle possesses an electric dipole moment, then time-reversal invariance is violated by the basic laws of physics. Experimenters have also worked to see if the neutron has an electric dipole moment.

The search for the electric dipole moments of the electron and of the neutron has now been going on for almost forty years, pursued by a group of persistent and, dare I say it, heroic experimenters who over the years have exercised their ingenuity to improve the accuracy of their measurements. In this country, there are currently efforts to find an electric dipole moment at the University of Washington, Yale University, Amherst College, and the University of California at Berkeley.

The story of the search for an electric dipole moment in nature illustrates well the difference between theoretical and experimental physics. It is all too easy for theorists to sit around and imagine an electron precessing in an electric field; it is an entirely different matter for experimenters to think of a way actually to observe this. When Gene Commins, the leader of the experimental group at Berkeley, came to the university where I work to give a lecture on time reversal, he drew an appreciative laugh from the audience when he remarked that "what a theorist like Tony Zee thinks about for an afternoon will take me twenty years of hard work to prove or disprove."

For one thing, you can't just stick an electron in an electric field; the electric force exerted by the field on the electron will push the electron out of the experimental apparatus in an instant. The experiment is actually performed on an atom. An atom is, of course, an electrically neutral collection of electrons and a nucleus. If the electron has an electric dipole moment, then the atom will also have an electric dipole moment. Over the years, experimenters have had to overcome hordes of technical problems, not to mention problems caused by increasingly meager funding from Washington. Commins and his friends can always dream of private funding. Hey, what if we lived in a society that wanted to be enlightened perhaps a teeny-weeny bit as much as it wants to be entertained? What if a tiny fraction of the profits from the movies on time reversal could be diverted to research on time reversal? Does anybody know Michael J. Fox's phone number?

After all this, I should tell you what the electric dipole moment is measured in—in centimeters, in grams, or what. Let us picture an electron as a tiny ball. Once again, think of our childhood toy the top, with that lopsided shape tapering down to a point. The lopsided shape is necessary; otherwise the top won't precess. No child has ever played with a perfectly round top. Thus the electric dipole moment of the electron, roughly speaking, is a measure of the extent to which the electron may not be perfectly round. Picture the electron as ever so slightly egg shaped. Then the difference between the lengths of the long and short axes measure the lopsidedness. The electric dipole moment is measured in units of length such as centimeters.

Now I can give you an idea of how hard experimenters have worked over the years. Back in 1959, experimenters were able to say that the electric dipole moment of the electron is smaller than 10^{-15} centimeters. In some thirty years, the measurement has improved to 10^{-27} centimeters. In other words, if the electron is lopsided and not perfectly round, its lopsidedness is less than 0.000000000000000000000000001 centimeters, by far the smallest length ever measured by scientists. Personally, I am amazed by what experimenters can do.

Meanwhile, the theory of time reversal remains rather primitive. It is easy enough for theorists to include an arrow of time in various theories of the fundamental interactions. Several years ago, when I visited Norval Fortson, an experimental physicist who has spent years looking for the electric dipole moment, in his lab at the University of Washington, I decided to cheer him up by concocting on the spot a theory in which the electron had a large electric dipole moment, big enough so Norval couldn't help but see it soon.

There are plenty of theories of time-reversal violation, but theorists often don't even realize that some of those theories might contain large electric dipole moments. A couple of years ago, the distinguished theorist Steve Weinberg came to the university where I do my research and told a gathering of physicists that according to a certain theory the neutron might have a rather large electric dipole moment. This came as a surprise and an embarrassment to the audience; the theory had been kicking around for some ten years, and all kinds of brilliant types had worked it over without seeing the possibility of a large electric dipole moment. Immediately after Weinberg's talk, one of my colleagues, Steve Barr, and I realized that the electron would also have a large electric dipole moment, perhaps not much less than 10^{-27} centimeter.

The irony surely did not escape you: what the experimenters call an incredibly small length that will take them years to measure, the theorists refer to as large. Here you have one picture of the sociology of physics. A bunch of guys painstakingly spend their lives rummaging through the haystack looking for the proverbial needle. Some other guys are hanging around on the sideline, once in a while going through a cheerleading routine: "Come on, guys, just look a little harder!" Back comes the reply, "We can't find it! We've looked for more than thirty years! If it's here, it's got to be shorter than 10^{-27} centimeters." "Hmm"—the guys on the sideline scratch their heads and then yell, "Hey, according to this calculation here, it just might be a wee bit shorter than 10^{-27} centimeters. Don't give up!"

What makes time-reversal invariance even more mysterious is that since 1957 physicists have known that the laws of physics do not respect space reversal. As time reversal interchanges past and future, space reversal interchanges left and right. Indeed, every time we look in a mirror we have arranged for space reversal, known as *parity* to physicists. What you do is show a bunch of physicists a movie reflected in a mirror. In 1957, physicists figured out how to tell that they are seeing the movie in a mirror. So why can't they tell whether the movie is being played backward or not?

Another clue comes from mathematical reasoning. A mathematical theorem called the CPT theorem asserts that the laws of physics, in the way that physicists know how to formulate them, must be invariant under the combined operations of space reversal, time reversal, and something called charge conjugation (which I won't go into here). For example, if we formulate a law that respects charge conjugation but violates space reversal, then it must violate time reversal. In 1964, experimental physicists found that the disintegration of a subnuclear particle called the *kaon* violates space reversal and charge conjugation in such a way that, according to the mathematical theorem, the disintegration must also violate time reversal. For more than a quarter of century, then, physicists have had indirect evidence that the laws of physics indeed violate time-reversal invariance. What physicists have been searching for in vain is direct evidence that time-reversal invariance is violated, without appealing to some fancy mathematical theorem. Physicists are a real skeptical lot and are never completely happy with mathematical theorems. Mathematical reasoning in itself cannot be wrong, but theorems have to start somewhere, with assumptions; and the assumptions that go into the proof of the CPT theorem, while seemingly reasonable to the vast majority of physicists, still make some physicists sit up at night and sweat.

To summarize the mystery, we know the butler committed similar misdeeds before: he did something to a mirror once. We sure suspect the butler, but we just can't find the evidence that he did it. We even have an expert—the guy proved a mathematical theorem—swearing that the butler must have done it, but the judge won't admit the expert's testimony in court. And so those detectives just have to keep on combing the rug.

Finally, I should mention that time reversal is what physicists call a symmetry in the laws of physics. The symmetry between left and right, or symmetry by reflection in a mirror, is of course the most familiar, beloved by artists and architects. Time reversal is a symmetry between past and future. When I was writing my popular-physics book *Fearful Symmetry,* I put off time reversal until the last chapter because I really didn't, and still don't, understand the nature of time. Neither does anybody else.

Early in this century, Einstein told us that space and time are intimately related. Yet, as a conscious being, I know damn well that I am free to move left or right but just can't move into the past when I please. If I didn't know any better, I would have thought that the laws of physics, while perhaps symmetric between left and right, would not be symmetric between past and future. Experimenters have found precisely the opposite: the laws know about the difference between left and right, but not between past and future.

There, I've mentioned the words so dreaded by physical scientists, *conscious being*. Time is the one concept in physics we can't talk about without dragging in, at some level, consciousness.

"Well, when and if Gene Commins, Norval Fortson, and their friends see the electric dipole moment, will they be able to tell us how to build that time machine?" you ask. Almost certainly not, at least not right away. The discovery of time-reversal violation in the fundamental laws may well represent a crucial step in the unraveling of the mystery of time. Who knows where it would lead eventually? Logically, however, the discovery of time-reversal violation in the fundamental laws and the construction of a time machine are not necessarily connected in any way. Think of the case of space reversal. We can imagine a civilization on some planet without water, without mirrors. Conceivably, physicists in this civilization could eventually have discovered that the fundamental laws of physics are not invariant under space reversal, but this still would not tell them how to manufacture a mirror.

Sorry to disappoint you, but physicists haven't yet figured out the difference between past and future, and they certainly can't send you on a ride with Michael J. Fox anytime soon. In fact, a physicist long ago pointed out a logical paradox against traveling backward in time known as the grandfather paradox. Indeed, all those popular movies about time travel basically play on that paradox. If Fox couldn't punch out the bully and help his father get a date, where would he be? If Michael Biehn couldn't save the mother of the rebel leader from the Terminator, what would have become of him? You can always invoke the notion of parallel universes, with reality branching off at every instant into alternate realities, but that's too wild for me. Hey, I'm a physicist, not a metaphysicist.

Go back to that little picture of how physics is done. Over there we see the haystack. Where are the people writing popular articles and books about physics? They are somewhere in left field, hawking popcorn to the

—

masses. Like everything else, popular books on physics span the whole spectrum: the good, the bad, and the ugly. To keep on selling, some people feel they have to keep on hyping. I go to the bookstore and see books about parallel universes, the universe before the Big Bang, the universe before time, and I cringe. Let's not forget the real heroes out there burrowing in the hay, ruining their eyes and fighting the mice.

As a university professor, I feel duty bound to include at least a few footnotes.

* As a student, I learned about time reversal from J. J. Sakurai, *Invariance Principles and Elementary Particles*, (Princeton, New Jersey: Princeton University Press, 1964).

* That the existence of an electric dipole moment indicates a violation of time-reversal invariance was pointed out in the 1950s, with varying degrees of clarity, by E. Purcell, N. Ramsey, L. Landau, C. N. Yang, and T. D. Lee.

* Some of you may have realized that I pulled a fast one on you. When we play the movie backward for an electron precessing in a magnetic field, the magnetic field reverses direction, but for the top precessing in the earth's gravitational field, the gravitational field certainly doesn't reverse direction. Yet both phenomena are time reversal-invariant. You may be wondering why there is a difference. The subtle difference is that the top, but not the electron, has a geometric axis.

DOES ANYBODY KNOW THE RIGHT TIME?

ROBERT H. MARCH

Arecibo, on the sunny northern coast of Puerto Rico, is just the ticket for a January getaway. And when the standard tourist diversions become tiresome, just a short drive from town is a most peculiar landmark, a snug valley that looks for all the world like a giant colander.

Spanning a bowl-shaped valley is a web of steel mesh 1,000 feet in diameter tethered to cables that hold it in a nearly perfect parabolic shape. It is there to gather in faint radio signals from outer space and bring them to a focus, for this is the "mirror" of the world's largest radio telescope. Unlike most other telescopes, it cannot be aimed at all. Instead it can only stare fixedly straight up, waiting for objects in a narrow band of the sky to pass within its view as the Earth slowly turns.

A pair of binoculars might reveal that high above the net a sensitive radio receiver is being hauled into position by taut steel cables. For in a few minutes, the "Best Clock in the Universe," the pulsar PSR 1937 + 21, will be passing overhead.

When it comes into range the pulsar will be unmistakable, emitting sharp "ticks" of radio noise 642 times each second. These signals will be fed to a computer and checked against the telescope's atomic clock, which will in turn be compared to other clocks passing overhead in satellites and, through a series of such comparisons, eventually to the "Best Clock on Earth," which is in Paris, of course. Way back in Napoleonic times the French appointed themselves the guardians of the world's standards of measurement, and nobody has seriously challenged their claim to this role, even though the standards themselves are now governed by an international commission.

Comparing all of these clocks takes a great deal of careful computation, and weeks or months will pass before the head of this project, Princeton astronomer Joe Taylor, knows how the reading of his "star clock" compares to those of clocks on our own planet. For starters, that Best Clock on Earth is a fiction, actually a table of clock corrections issued long after the time they apply to. A computer at the International Bureau of Weights and Measures compiles this table by comparing the readings of more than a hundred master clocks scattered around the globe. But one thing is certain, both from theory and from experience: since it is January, *every good clock on Earth is running slow.*

Albert Einstein foresaw this slowdown more than three-quarters of a century ago. It originates in the combined effects of the Sun's gravity and the Earth's orbital motion, both of which tend to slow down time. Early in January, the Earth is about 3 percent closer to the Sun than in July and is moving at it fastest, so both effects are maximized. But it was not until the 1980s that clocks were found outside the Solar System that were accurate enough to give the prediction a meaningful test.

Of all of the billions of people on Earth, only a handful of astronomers have any reason to care about this discrepancy. In ordinary terms it is a tiny effect. By early April, Earth time will fall only 1.6 milliseconds (thousandths of a second) behind star time. It will then begin to catch up, pulling even by July and creeping 1.6 milliseconds ahead by October. Since the smallest time interval noticeable to our senses is about 20 milliseconds, this is hardly something for most of us to get excited about.

The 1.6 milliseconds is just a seasonal variation, owing to the Earth's orbit not being a perfect circle. All Earth clocks run more slowly than the star clock all of the time, losing nearly 0.5 seconds each year. But this is a constant factor, so it is more convenient to measure the star clock's pulses against the mean rate of an Earth clock averaged over a whole year. An Earth second is defined as 9,192,631,770 pulsations of a cesium-beam atomic clock. At least so far, nobody out there has raised any objections. Average Earth time, officially known as TDB (for *bary-centric dynamic time,* using French word order) is the astronomers' time scale. It is the time that would be kept by a hypothetical cesium clock at sea level on a fictitious Earth that swings around the Sun at a constant speed in a perfect circle, in an otherwise empty Solar System. (The arcane world of precision timekeeping is populated by nearly as many hypothetical clocks as real ones!) This time scale is obtained by correcting real clock readings for the seasonal variation, plus a smaller correction for the effect of the Moon on the Earth's motion.

Einstein tells us that it is not just our clocks but *time itself* that is affected. Absolutely nothing here on Earth will seem the least bit different, be it January or July, for everything that happens will be slowed by exactly the same amount as the clock that is timing it, including our own biological sense of time. It is only when we check the precise timing of things happening far from our planet that we are reminded that our clocks have lost a few milliseconds.

Thus the rest of us might be inclined to ask, "Why bother?" The answer is that for modern technology, it is sometimes not just milliseconds but fractions of *micro*seconds (millionths of a second) that count. And in eight years of watching his sky clock, Joe Taylor has demonstrated that at that level, the best Earth clocks are *not nearly as good as we would like them to be.*

When Einstein formulated his theory of relativity early in the twentieth century, it was an exercise in the purest form of abstract science. But as the century draws to a close, it has become part and parcel of the workaday world. Indeed, Einstein's most mystifying precept, the assertion that time may be considered a fourth dimension, has even become part of that most mundane branch of practical knowledge, our system of weights and measures. (To set the record straight, Hermann Minkowski, one of Einstein's teachers, is the true father of the theory of the fourth dimension, though he formulated it as an interpretation of Einstein's work.)

On its most basic level, the theory of the fourth dimension is simply the idea that, as every traveler knows, time and distance are connected.

For anything traveling at 299,792,458 meters per second (the speed of light and the official limit to all possible speeds), a small time can correspond to a rather large distance.

When I call the speed of light "official," I mean precisely that: its value is now set by international agreement. This is because since the 1960s, we have been able to measure time far more accurately than we can measure distance. The meter is no longer defined, as it once was, by two marks on a platinum-iridium bar, or as it later was by so many wavelengths of light emitted by the element krypton. The meter now is simply "the distance light travels in $\frac{1}{299,792,458}$ of a second." This means that there is no longer any reason to measure the speed of light. Our standard for time is now also the basis for our standard of distance. Any measurement that disagreed with the official value would simply mean that the experimenter did not measure meters properly.

And the speed of light is certainly a "limit," notwithstanding anything you may have heard about *tachyons,* particles that move faster than light. Even if these hypothetical (and, after many failed attempts to find them, rather improbable) objects exist, the speed of light is still an impenetrable barrier, with ordinary matter limited to slower speeds and tachyons fated always to move at faster speeds.

The speed of light is not simply a *formal* standard but also a *practical* one. Our best navigational aid today, the Global Positioning System (GPS), depends on radio signals traveling at the speed of light. They are transmitted from orbiting satellites, and they tell us what time it is and where the satellite is located at that instant. If you know what the time is when you receive the signal, you know how far you are from the satellite. If four GPS satellites are in range of the receiver, a computer can solve for the receiver's position and the correct time.

Since light covers about 300 meters, about one thousand feet, in a microsecond, an error of 1 microsecond in time means an error of 300 meters in location. While any old salt can tell you that this is awfully good by the standards of old-fashioned celestial navigation, it is not nearly good enough for a lot of things the GPS is supposed to do.

Like so many American technical achievements of the Cold War era, the GPS was developed for military use. On the battlefields of Desert Storm, which were often devoid of obvious landmarks, the only way to be sure whether some indistinct blob in an infrared gun sight was one of "our" tanks or one of "theirs" was to call out on the radio and ask all the "good guys" to report their positions. The most accurate part of the GPS time signal was supposed to be scrambled, so that only our military could take advantage of the full design accuracy of 10 meters. Civilians were only supposed to be able to estimate positions to give-or-take 100 meters. But the Gulf War caught the U.S. forces without enough military GPS receivers to go around, so the Pentagon turned off the scrambler and bought up thousands of civilian receivers.

As you might imagine from the recent history of consumer electronics, these off-the-shelf models turned out to be easier to operate and just as reliable as their military counterparts, at a fraction of the cost. But despite the end of both the Gulf War and the Cold War, the military loves a secret, so within a year scrambling was resumed.

In any event, civilian manufacturers have long since learned how to get around the codes, at least for a receiver that isn't moving very fast. The GPS can provide survey data so accurate that geologists can monitor the slow grind of continental drift, a matter of inches per year, on a month-to-month basis.

For most of recorded history, however, our standards of measurement and transmission of time were based not on the rapid beats of atomic vibrations and the speed of light, but on the Earth's far more leisurely rotation. Though today a sundial is likely to serve only as a lawn ornament, for many purposes it is still a pretty good clock.

The sundial, too, is affected by the changes of speed as the Earth traces its path around the Sun. A solar day is the time between solar noon (Sun at its highest point in the sky) one day and the next, and it is a few seconds longer in January than in July because when the Earth is moving most rapidly it must turn a bit farther to face the Sun. But the day-to-day change is gradual, and a sundial will never stray from a steadier clock by more than 16 minutes. Since this variation is small compared to the typical seasonal variations in sunrise and sunset, solar time would be good enough for the way most of us regulate our lives even today. Of course, Sun time also differs from place to place as the Earth turns, but in the absence of high speed travel or communications, there was no need for everyone to keep the same time.

With the advent of the astronomical telescope, a system was set up to use the stars rather than the Sun as the marker for the Earth's rotation. The time between passages of a star in the cross hairs of a fixed telescope remains steady throughout the year, about 4 minutes less than an average solar day. Since the Sun still remains the timekeeper for the rhythms of life, the astronomers who maintain our time scales tacked on these 4 minutes, creating the world's first hypothetical clock, Greenwich Mean Time. Like TDB, GMT is the time that would be kept on a fictitious Earth in a circular orbit around the Sun by a miraculous 24-hour sundial located in Greenwich, England. This standard of time endured right into our own century.

Just as today, the technology that drove precision time-keeping was that of navigation. Determining latitude—the distance from the equator—has never been a problem. In the northern hemisphere, it is simply the elevation of the North Star, Polaris, above the horizon. But to use the stars to find longitude at sea, a navigator needs a clock that indicates, even after a long sea voyage, what time it is in Greenwich. Since the Earth turns about 15 degrees each hour, the star pattern we see now will be seen 15 degrees to the west an hour later. One need simply record the time when a star reaches a certain position in the sky and consult a table showing when this happened at Greenwich. Every minute of difference corresponds to one quarter degree of longitude— about 17 miles at the equator and less at higher latitudes. Preparing and updating the navigational tables became a major contribution of Green- wich and other national observatories, and through this service, astron- omy, that least practical of all sciences, has certainly paid for itself many times over.

In 1713, the British government decided that a clock that could keep time to 2 minutes after several months at sea would be good enough, and they offered a prize of the then-princely sum of twenty thousand pounds to the first clock maker to attain this level of perfection. The best existing clocks were based on the swing of a pendulum, which rendered them useless on a ship buffeted by winds and waves. The quest proved quite a challenge, and many skilled artisans were left by the wayside. It took John Harrison, the self-educated son of a Yorkshire carpenter, over 40 years to perfect a clock that could claim the prize. It then took him 5 more years to receive the full promised stipend.

Harrison's chronometers, with continual improvements, were to rule the sea well into the twentieth century, when they were supplanted by radio time signals broadcast direct from Greenwich and the world's other standards observatories. If you live in the United States and have a short-wave radio, you can get the right time from radio station WWV in Fort Collins, Colorado, or station WWVH on the island of Kauai, Hawaii. Both broadcast at 5, 10, 15, and 20 megahertz. If the voice you hear on the time announcements is feminine, you're listening to Hawaii.

But even as Harrison was perfecting his marvelous clock, the philosopher Immanuel Kant was pointing out that, ultimately, the Earth could never be the last word in time-keeping because its rotation is slowing imperceptibly due to tidal friction. The gravity of the Moon and Sun raises a tidal bulge a few feet high in the open sea, and this tidal bulge moves from east to west. When this wave passes through shallow seas or collides with the land, the drag it exerts retards the rotation of the Earth. As a result, a day in the year 2000 will be about 20 milliseconds shorter than a day in 1900.

For more than 200 years, tidal friction remained a conjecture. The pendulum clocks used by the world's standards observatories right through the 1950s could scarcely be trusted to be accurate within 20 milliseconds from one day to the next. It was in that decade that the first clock to keep better time than the Earth—the quartz-crystal oscillator—came into widespread use. Not only could it actually measure the slowing of the Earth's rotation, but it showed that this slowing is not a steady process because the Earth is not a static body.

One of the sources of variation is seasonal: when it is winter in the northern hemisphere, which contains most of the world's high mountain ranges, deep snow packs form at high altitudes. This movement of mass upward from sea level slows the Earth's rotation, just as a dancer or skater can slow a spin by extending his or her arms. With the spring melt-off, water runs back to the sea and rotation speeds up again. One particularly long or snowy winter can retard the Earth by tenths of a second. Geological processes and flows deep within the Earth contribute irregularities on a somewhat longer time scale.

Today these wonderful quartz clocks have been miniaturized, with the result that a twenty-five-dollar quartz Timex watch rivals for accuracy a mechanical Rolex costing thousands more. But good as they are, quartz crystals never had the chance to serve as *the* time standard. For one thing, the frequency of a quartz crystal is not a natural constant, like that of cesium atoms; it depends on the size of the particular crystal. And as quartz clocks were being perfected, the far more precise cesium clock was not far behind, promising an accuracy of nanoseconds (billionths of a second) per day.

Nonetheless, there were still good and proper reasons to keep the Earth as the official time standard. In the 1960s, celestial navigation was still widely used and still demanded a clock firmly tied to the Earth's rotation. There was also a strong feeling that sunrise and sunset ought to stay put over the years, rather than drifting later by a minute or so per century. So from 1958 to 1967, the keepers of the world's time standards resorted to the desperate expedient of the "rubber second." Each year, the experts in Paris changed the length of the second to keep the mean solar day exactly 86,400 seconds long. But since crystal oscillators and atomic clocks are not easy to adjust and the slowdown is a bit unpredictable, in practical terms this solution turned out to be a terrible nuisance.

In 1967, scientists officially cut all ties with the Earth's rotation, while civil time-keeping stuck with it for a while. Then in 1972 a "hybrid" of the two systems became the standard for time broadcasts. The scientists' time, International Atomic Time (TAI, using French word order) follows the cesium clocks, oblivious to what the Earth may be doing. A second in civil time is the same length as one in TAI, and both time scales tick in unison. But to keep the new civil scale, known as Coordinated Universal Time (UTC), in step with the Earth, the pundits in Paris add a "leap second" from time to time, either at the beginning of the new year or the start of July. By the end of 1992, TAI was exactly 26 seconds ahead of UTC.

With this scheme, scientists can take UTC from the radio signals, add the leap seconds, and get TAI. Those who still steer by the stars can navigate on UTC to an accuracy of less than 1 second, or better than 0.25 of a mile. And for those who prefer the old GMT standard, now known as *UT1,* it is still there on the time broadcasts in coded form.

With the adoption of the atomic time scale, the world's timekeepers suddenly faced a new problem: how to be sure that all of the world's standard clocks were keeping the same time. The old system of shortwave radio had been good enough for GMT, which was only reliable to about 0.10 seconds. For UTC, however, shortwave just wouldn't do, because long-distance transmission does not go as the crow flies. It "skips" around the globe, bouncing back and forth between Earth's surface and the ionosphere, the upper layer of the atmosphere. The altitude of the ionosphere changes from hour to hour, so it is hard to estimate how long the signal is in transit to much better than a millisecond.

The solution, for the 1970s and most of the 1980s, was *portable* atomic clocks, flown from one observatory to another on regular passenger flights. They couldn't simply be shipped air freight, for the early portable clocks were touchy devices that needed a human traveling companion to cater to their whims. It was a nice job for a young engineer who wanted to see the world, and frequent flights kept most of the world in step at the level of a few microseconds. It was this procedure that first brought relativity into our system of weights and measures: the speed and altitude of the plane have significant effects on the time kept by portable clocks, and a careful flight log must be kept in order to calculate the required corrections.

This era is coming to an end with the development of atomic clocks that can run unattended for years in orbiting satellites. Each GPS satellite carries four clocks that are constantly checked against each other, and at frequent intervals against master clocks at the Air Force Academy in Colorado Springs, Colorado. These are closely linked to the official U.S. standard clocks at the National Institute of Standards and Technology in Boulder, which keeps in touch with a second U.S. standard at the Naval Observatory in Washington, D.C. The Defense Department insisted that it had to keep its own time laboratory when the civilian one was created.

These standard clocks, by the way, are not the sort of thing you could set on your mantelpiece—the heart of the device is a beam of cesium atoms in a 30-foot-long tube. If it's just accurate time you want, your best bet is a GPS receiver, which will soon cost less than a thousand dollars.

If two laboratories are simultaneously listening to the same GPS satellite, they should in principle be able to synchronize their clocks to within 0.01 of a microsecond. Such comparisons go on all through the year, between pairs of labs that are close enough to permit it, and these measurements are combined to form the "Best Clock" average. But when this is compared to Joe Taylor's star clock, the result is unsteady, varying at the level of a few tenths of a microsecond from one observation to the next. Paradoxically, some individual clocks seem to do better than this grand world average, but this is not of much use for comparing data between far-flung astronomical observatories.

Taylor suspects that the fault lies in the procedures for synchronizing Earth clocks, for PSR 1937 + 21 is an exceptionally stable object. It is a solitary *neutron star,* essentially a giant atomic nucleus 50 percent heavier than the Sun but only a few miles in diameter. It sits alone in space, far from the gravitational influence of any other object, and acts as a massive flywheel. It whirls about as fast as such an object can without tearing itself apart from centrifugal force, and it sends out a radio beacon that sweeps past the Earth at each turn. It is hard to imagine how this rotation could be anything but steady.

The "time lords" in Paris think they understand at least some of the sources of the jitter, and they are working to get rid of it. Many laboratories are searching for something steadier than the cesium standard. In the meantime, Taylor and others are studying a number of alternative sky clocks, to see how well they compare with each other. Perhaps they could provide the benchmark against which future improvements in time-keeping can be tested.

By the mid-1990s the GPS should have twenty-one satellites in orbit, giving 24-hour coverage for the entire globe. Around Orlando, Florida, General Motors is testing a GPS navigation system that shows a car's position on a video map display. The Federal Aviation Administration and its foreign counterparts would like to make GPS the standard for global air-traffic control, because at its full design accuracy GPS should be able to steer a plane right to the threshold of any runway in the world, provided of course that the Pentagon can be persuaded to give up on scrambling.

Thus, someday soon the safety of the air traveler may, like that of the seafarer in John Harrison's day, rest once again on the knowledge of astronomers and the skills of clock makers.

How Memory Speaks

JOHN KOTRE

It's a great puzzle, this memory of ours. A scene we experience for a moment—and only once—remains clear in our minds for a lifetime, yet we forget the looks of things we see and touch almost every day. Once, when I recorded the life story of a young woman, she spoke of a glance that never left her. She was recalling the birth of her first child: "I remember taking him to my mom—my mom had not seen him—and I'll never forget the look. It—she—just crushed me. She looked at him as if to say, 'Is that him?' "

There are moments, good as well as bad, that none of us will ever forget. But there are plenty of ordinary objects and events for which our memory is a blur at best. How much, for example, can you remember about a penny? A lot, you say? Well, try drawing one—now.

It ought to be easy, but it isn't. When you get stuck, compare what you've drawn with an actual penny. On the "heads" side, did you have Lincoln facing toward the right, where the year 19—appears? Did you have In God We Trust written across the top and Liberty on the left? For "tails," did you draw a building in the middle (it's the Lincoln Memorial), and did you write United States of America across the top, E. Pluribus Unum just below that, and One Cent across the bottom?

The difference between a mother's momentary glance and the details of a penny is, of course, the meaning of each. In a study by psychologists Raymond Nickerson and Marilyn Jager Adams, only 20 percent of the subjects recalled even half of a penny's 8 features correctly. But one subject got everything right. He happened to be an avid coin collector;

he could picture pennies in minute detail because they had a special place in the story of his life. They were part of what researchers are now calling *autobiographical memory*.

Autobiographical memory is memory for the events and feelings that go into the story of a life. For many years, clinical psychologists, psychiatrists, and social workers have dealt with this kind of memory. It's the stuff of which "insight" therapies and "talking" cures are made. I have worked with autobiographical memory, too, but in a different context. I have recorded the stories of people's lives and put them—or parts of them—into books, audio programs, and on one occasion a public television series called "Seasons of Life." And I've wondered again and again about the mystery of memory, about mental pictures from long ago that warm the face of one storyteller, moisten the eyes of another, and bring the "shakes" to a third. Are these memories photographic or even remotely accurate? And, accurate or not, what do they mean? What do the memories speak of, and how do they speak?

Thankfully, a new breed of memory researchers ventured forth from the laboratory in the late 1970s in pursuit of "ecological validity." Abandoning research on nonsense syllables and word lists—on memories of a few minutes' duration—they began to investigate such real-world phenomena as legal testimony, diaries, and recollections of historical events. Elsewhere in psychology there grew a new interest in narrative forms of thought. Taken together, these approaches are helping us decipher the special language of autobiographical memory, helping us understand the stories we tell ourselves about ourselves, the stories we'll eventually pass on to our successors.

WHAT PART OF OUR LIVES DO WE REMEMBER BEST?

A man I interviewed some years ago—I'll call him Chris Vitullo—had a vivid, 70-year-old memory of two coins. Chris looked like a clean-shaven Santa Claus. He had neatly trimmed white hair, a red nose, a thick torso that rested on sturdy legs, and a hoarse Sicilian accent. The coins he remembered weren't pennies, and Chris would have had a hard time recalling the features of each. But they did have an important place in the story of his life.

Chris spent his first four years in Sicily while his father was working in America. But the day came when his father rejoined his family. "I'm getting the goose pimple now," Chris said as he relived the experience of meeting him for the first time. His mother had told him what to do. "I kissed his hand, and I asked him to bless me. *'Mi benedica, Padre.'* Naturally, my dad, he picked me up and he put me in his chest, and then he put two fingers in his vest pocket and he got a couple big coins, silver coins. They were dollars, and he gave them to me. Just think how tickled I was, huh? Oh, boy! The first time I met him, he gave me two coins made out of silver, two silver dollars."

This is one of a host of memories that Chris produced from his childhood, adolescent, and young-adult years. The sheer number of memories raises an interesting question about autobiographical memory: what part of our lives do we remember best? We can't possibly recall everything that happened to us—every hour of every day of every year. When we think of our life as a whole, to what stages do our thoughts instinctively turn?

Chris Vitullo's thoughts turned to his beginnings. He spoke of the tales his mother told him as a child, of his apprenticeship at the age of 7 to an old barber named Antonino, of his voyage, alone, to America when he was 13. Leaving his mother and father on the dock, "I had a lump in my throat as big as a fist." He remembered everything about the trip: the strange languages and the strange foods, the way the boat listed so that Chris could see the ocean on one side but not on the other. Then someone spotted the Statue of Liberty. *"La liberta! Viva la liberta!"* From New York, Chris took a two-day train trip to Saint Louis, got off at the wrong station, and then, with an incredible bit of luck, walked straight to the front door of his sister's home. All these events Chris recounted in great detail. Several years after arriving in Saint Louis, he moved to

Detroit and opened a barbershop with a friend. His father and mother came from Sicily to arrange his marriage to a girl named Gloria, and by the age of 21 the course of Chris's adult life had been set.

Chris was full of energy as he described these events. But once he came to his marriage to Gloria, his narrative slowed to a crawl. "When we got married," he said, "we were very happy after that, with an exception that whatever comes along, you have to take, sickness or otherwise. We lived together forty-five years, Gloria and me. Yeah, we lived together forty-five years. We had a lot of good times together and a lot of bad times together. We worked, we paid our bills honorably, and we dressed well. We bought a house, we paid for the house, we had a little bit of money, whatever God provided, but we did it all in a good faith and honestly, and we arrived to the point that probably if God wouldn't want Gloria to pass away, maybe Gloria and I, we would be together today. But that's the way it goes in life, and we have to take what's coming to us."

There it was: after hours on his first 21 years, a minute or two on his next 45. In subsequent interviews I was able to learn more about Chris's marriage to Gloria and her death 10 years prior to my first interview, but it took a lot of questioning and checking. Chris wasn't resistant to speaking of this time; it was just that his memory had condensed nearly half a century into a single whole, with much of the detail forgotten. What Chris remembered best—and with the greatest joy—were the years of his life from 6 to 21.

There are quicker ways to pinpoint the years that stand out in most of our minds than gathering individual life histories. One of the simplest goes back over a hundred years to Sir Francis Galton, who is best known for his innovations in statistical analysis and his obsession with measuring psychological traits. Galton's memory technique is akin to Freudian free association. Subjects are given a cue word—*window,* for example. They then report a memory that the word triggers. After several dozen cues are presented, subjects go back and date each remembered incident as closely as possible. You can try it yourself. Common cue words are *avenue, box, coin, flower, game, mountain, picture, storm, ticket,* and *yard.*

When memories are dated in this fashion and averaged across a number of subjects, an interesting pattern emerges: People in the second half of life do not apportion their memories the way people in the first half do. Memory researcher David Rubin has pieced together the results of half a dozen studies and found that in response to cue words, 20-year-olds produce a high proportion of very recent memories and relatively few from the distant past. Thirty-year-olds aren't much different, once you allow for the fact that they are looking back on an additional decade of life. In general, the responses of both age groups fit the normal forgetting curve. This curve looks like a slope for expert skiers. Most of our memories are of recent events—particularly those of the last few days. Then there is a rapid decline in retention until a point is reached at which forgetting becomes more gradual: we've lost most of what we're going to lose. At this point, the forgetting curve turns into a gentle slope that carries us back to the beginning of life.

But not to the very beginning. When you're dealing with autobiographical memory, the normal forgetting curve runs into a drop-off that wipes out recollections of our earliest years. People vary a great deal, but most say their first memory of life comes from when they were 3 or 4. (Women generally report an earlier date for their first memory than do men; the average difference is several months.) In our first 3 years, our brain isn't mature enough, nor is language developed enough, to enable us to store episodes in memory. The darkness covering these years is known as infantile amnesia.

All age groups experience the drop-off of infantile amnesia. But only in the second half of life does something else happen. Responding to the same cue words as younger people, subjects in their 50s and 70s (the only other age groups for which data are available) report a disproportionate number of memories from the early years, especially the second and third decades of life. This bonus of memories forms an extra hill at the bottom of the forgetting slope, just before the drop-off of infantile amnesia. Rubin calls the hill *reminiscence,* something over and above normal remembering. We'll return later to the question of what it means.

The cue word method for dating autobiographical memories may seem a bit artificial, and a lot depends on how you set up the experiment, but I have found that the results fit a surprising number of life-story tellers. My experience has been that the hill begins even before the second decade of life, around the age of 6, when a string of continuous episodes—not just isolated fragments—can first be found in adults' memories.

At 76, Chris Vitullo fit the basic profile of 70-year-olds, though his excess of memories came from the second decade of his life rather than the second and the third. Remarkably, his pattern was quite close to that of another 76-year-old—a woman who had lost a good deal of her memory and who, in the words of Mark Twain, could remember anything "whether it happened or not." Recording her story, I had little idea which of her memories were fact and which were fantasy. Yet the temporal location of her recollections fit the pattern of forgetting slope, reminiscence hill, and infantile drop-off. A phenomenon that persists even when you're dealing with fantasy makes you wonder what autobiographical memory is up to. Let's see what's going on.

A Keeper of Archives

Not that long ago, many psychologists thought of memory as a library that stored every experience we ever had, something like a video recorder that was always on. Freud's views on the repressed unconscious contributed to this model of memory; a lesser-known influence was the work of Montreal brain surgeon Wilder Penfield. In the 1940s, Penfield treated patients with severe seizures by removing damaged tissue from the outer layer of the brain, known as the *cortex*. While doing so, he was able to map certain areas of the cortex by seeing what body parts moved when those areas were electrically stimulated. Since the brain has no pain receptors, his patients were able to remain awake during the procedure. Occasionally they reported hearing things in response to stimulation, things like a mother calling a little boy. Penfield had a hunch that he was activating long-forgotten memories. They had always been there, in the brain, but inaccessible until touched by his electrode.

Hypnotists were also encountering fascinating experiences when they took people on age regressions. Under hypnosis, subjects were told that they were 3 or 4 years old; they started to talk like children and produce vivid memories of events they had apparently long since forgotten. Therapists saw in hypnosis a way of uncovering details of trauma in a client's past; criminal investigators saw a way to heighten the awareness of detail in witnesses. In 1949, Robert True published in *Science* magazine a study in which he took hypnotized volunteers back to Christmases and birthday parties at ages 4, 7, and 10. He then asked them what day of the week it was. The results were astonishing: 82 percent of the answers were correct. Had subjects simply been guessing, only 1 in 7 answers would have been right. Hypnosis appeared to be doing the same thing as Penfield's electrodes—finding the light switch in the darkest regions of the library of memory.

If you ask ordinary people what they think memory is like, most will reply with a version of the library model. At least, that's what psychologists who have done the studies have found out. People may picture memory as a storage chest or a tape recorder or a computer, but their basic belief is that everything—even the appearance of a penny—is *in there,* somehow, somewhere. The trick is finding it all. The vivid memories in which we have such confidence are simply the photographs that haven't faded with time.

In this view, memory is like a keeper of archives, a fastidious librarian who tries to keep original materials in pristine form. That, after all, is the ideal of memory. We are proudest of recollections that go back a long time and remain as fresh as on the day they were first filed away.

But the picture of autobiographical memory as a careful keeper of archives hasn't held up under recent scrutiny. When psychologists Elizabeth and Geoffrey Loftus examined the flashbacks of Penfield's patients more closely, many turned out to be fabrications. The patients recalled being in places they had never even visited, for example. And when researchers in hypnosis were unable to replicate True's age-regression results, it was discovered that he had not asked subjects, "What day of the week is it?" but rather, "Is it Sunday?", "Is it Monday?", and so on, and that he himself had known the correct answer when asking the question. That tiny detail made all the difference: it takes only the merest vocal inflection, even unintended, to communicate the right answer to a subject. What True had demonstrated was not the power of memory but the power of suggestion.

In the early 1970s, Elizabeth Loftus demonstrated the power of suggestion in the memories of nonhypnotized subjects. College students in her experiment watched a brief film in which a few seconds were devoted to a traffic accident. Afterward, some of the subjects were asked to estimate how fast the cars in the film were going when they "hit" each other. Others were asked to estimate the speeds when the cars "smashed" into each other. Both groups saw the same film, but the "smashed" group gave higher estimates of speed than the "hit" group. A week later, both groups were asked if they had seen any broken glass. More members of the "smashed" group, and more of those who had given high estimates of speed, said yes. Actually, there had been no broken glass in the film. Subsequent research has convinced Loftus that leading questions can change the mental picture of rememberers; they now *see* broken glass, or a white vehicle, or a stop sign, or a barn, or whatever has been suggested. Loftus believes that two kinds of information go into a memory. The first is the original perception of the event; the second is information supplied after the event. With time the two become blended into one "memory."

The word that's used for this blending is *reconstruction*. Memories don't sit inertly on the shelves of a library; they undergo constant revision. Think back to one of your earliest memories of life—a move to a new house, the birth of a baby brother or sister, a birthday party, an accident, whatever it might be. Do you see yourself in the memory? Many people do, and it's more evidence of the reconstructive nature of memory: when you originally experienced the remembered event, you weren't outside of your body looking at yourself. Several studies have shown that the older a memory is, the more likely it is to be rebuilt with an out-of-body observer.

If autobiographical memory is reconstructive, it turns out that the brain, our organ of memory, is no different—even in a physical sense.

Today we know what Penfield didn't, that in the course of brain development there is a continual weeding out of nerve cells and even of connections between cells. During the first half of the prenatal period, all the nerve cells that a brain will ever possess have already been born. At that point begins a massive die-off, which slows down but continues throughout the life span. Something similar happens with *synapses,* the connections that carry messages from one nerve cell to another. Their number reaches a peak around the age of 2 and diminishes after that. We know, too, that branches on nerve cells grow in complexity when one's environment is stimulating; if the environment is impoverished, branches fail to develop. In short, the brain revises itself over and over in the course of a lifetime. It's far from a static keeper of archives.

A Maker of Myth

One way of seeing how autobiographical memory reconstructs the past is by comparing a memory when it first enters the mind with the same memory years later. In 1972, psychologist Marigold Linton set herself precisely that task. Every day she wrote down on cards brief descriptions of at least 2 events from her day, 1 event per card. After a while she began giving herself monthly tests. Could she remember an event well enough to date it? At the end of six years, Linton wrote in her article "Transformations of Memory in Everyday Life," she had recorded over 5,500 events and was spending 6 to 12 hours on her monthly tests. After 12 years, she had learned a great deal about autobiographical memory.

In one way, her memory functioned very much like a library, storing recent events on shelves marked New Books—things that happened last week or last month. Linton was able to retrieve these memories with a simple chronological search. But after a year or so, events were moved to the main stacks of her memory, organized now in terms of their content—as things done with friends, for example, or things done in connection with work. Except for major landmarks, *when* faded as a retrieval cue; in comparison, *what* grew stronger. You yourself can probably remember what you did this past summer; but you would have a hard time recalling the events of three or four summers ago—unless you thought of them, say, as vacations or projects or episodes in a faltering relationship.

The fading of *when* is important because it leads us to autobiographical memory's real interest: the creation of a myth about the self. A *myth,* in the sense that I'm using the term, is not a falsehood but a comprehensive view of reality. It's a story that speaks to the heart as well as the mind, seeking to generate conviction about what it thinks is true. We think of myths as belonging to a culture—to a group. But there are also personal myths. When a myth is personal, it seeks to generate conviction about the self—about who "I" am.

One way memory makes myth is by deciding what an episode means and therefore where it belongs in the library. Interestingly, once the decision has been made, we no longer need to remember similar episodes. In her self-study, Linton was surprised to discover how much she had forgotten because events had lost their distinctiveness. She remembered a new class she taught but not all the times she had taught an old one, a match with a new racquet partner but not all her matches with a former one. What remained in her memory were unique events, the first times, but not all the subsequent times. Most libraries are interested in duplicates, but most memories are not. Duplicates contribute nothing to meaning.

Starting with the 4th year of her testing, Linton began to notice something else. A few of the cards that were supposed to jog her memory not only failed to do so but made absolutely no sense. She simply could not understand what she had once written. "I could hear my voice describe fragments from my own life that were somehow completely meaningless." The problem wasn't in the original writing. The events, rather, connected with no pattern that had developed in her life. They hadn't led anywhere, didn't fit anywhere. They were orphans in an autobiographical memory system fashioning generic memories of *what* events mean.

Psychologists have all sorts of names for generic memories: scripts, schemas, MOPs (memory organization packets), and TOPs (thematic organization points), to list just a few. No matter what generic memories are called, the idea is that we create them from the specifics of everyday life and arrange them in a kind of hierarchy. At the top of the hierarchy is a self, a person who says, "This is me, and this is how I got to be the way I am." Listen again to Chris Vitullo's generic memory of 45 years with Gloria: "We had a lot of good times together and a lot of bad times together . . . We had a little bit of money, whatever God provided, but we did it all in a good faith and honestly, and we arrived to the point that probably if God wouldn't want Gloria to pass away, maybe Gloria and I, we would be together today." Meaning is present in a myth of God's action, but nearly all the events that yield the meaning are absent. The condensation that has taken place is one of the ways autobiographical memory makes a long story short.

Generic memories have powerful effects. They can lead to the phenomenon known as *déjà vu*. You walk into a restaurant and have a strange feeling you've been there before. But you know you haven't. You've probably activated a generic memory—a "script" for entering restaurants. Studies have shown that generic memories can alter the recollection of specific details. A professor delivering a lecture may never point to information on the blackboard, for example, but many students will "remember" that she did. Pointing to the blackboard is part of the standard script for lectures.

These effects of generic memories are failures to meet the ideal of memory, to preserve the past in its original form. But something else is going on: the making of myth. Myth making is illustrated in a now-famous study of memory. In June of 1973, John Dean testified before the Senate committee investigating the Watergate cover-up about a meeting he had had with then-President Nixon nine months before, on September 15, 1972. Dean prefaced his testimony by saying that he believed he had an excellent memory. After hearing him testify, the press began to call him "a human tape recorder." When it was later revealed that an actual tape recorder had been recording during the meeting about which Dean testified, an experiment of nature was created. Ulric Neisser, whose *Memory Observed* opened up the ecological approach to memory, compared the transcript of Dean's testimony with the transcript of the actual tape-recorded meeting. He found rampant reconstruction. In a literal sense —who sat where, who said what—Dean's testimony wasn't even close to being accurate. But in the sense of what the meeting meant in the larger scheme of things, it was all true. Nixon had the knowledge Dean attributed to him; there was a cover-up. In Dean's mind, a single event—his meeting with the president—symbolized a pattern of repeated events. The symbol was so compelling that Dean "remembered" specifics that never took place on the occasion in question.

Dean's memory was inaccurate but true. It had decided on meaning and was making myth. Neisser's comparison showed that Dean inserted into his memory something he yearned for at the meeting but never received—an opening compliment from the president. In memory, Dean gave himself the benefit of hindsight, reversing a prediction that had proved to be wrong. And in his recollections he saw himself as more central to ongoing events than he really was. "What his testimony really describes," wrote Neisser in conclusion, "is not the September 15 meeting itself but his fantasy of it: the meeting as it should have been, so to speak. . . . By June, this fantasy had become the way Dean 'remembered' the meeting."

Autobiographical memory is interested in specific events, but only insofar as they contribute to meaning. Ultimately, meaning will arise in a comprehensive story of the self, a story replete with wishes and prophecies, a story that puts the self at center stage. By day, autobiographical memory may be a keeper of archives, but by night it's a maker of myth.

THE MOST IMPORTANT MYTH

As a psychologist, I've always been interested in dreams. Terrifying, comforting, amusing—no matter what their mood, they have such a sense of immediacy that it's often a jolt to leave them for the world of waking "reality." And the wisdom one sees in them . . . There are dreams from long ago that I never want to lose the memory of, so I repeat them to myself and to the most important person in my life. Sometimes I wonder if I could tell the story of my inner journey by touching on the dreams that came at the turning points, dreams that are mythic landmarks along the way.

But over the years I've discovered bits and pieces of mental life that are even more fascinating than dreams. They are fragments from our earliest years, the shards of remembrance that come just before the drop-off of infantile amnesia. What's the earliest thing in life you can remember? I've had an 11-year-old tell me it was a "dippy" dress her mother made her wear—and a profound sense of embarrassment. A strong young man of 20 said it was being in the hospital, where he fearfully awaited a shot from a nurse. An energetic woman in her 50s told me she remembered being bored and restless in a baby carriage; she wanted to escape the confinement. And a man in his 60s recalled that when he was 18 months old he walked to the edge of the front porch, only to be grabbed and pulled back by his mother. I've collected hundreds and hundreds of first memories, and, unlike dreams, they are all experienced as veridical—as the work of a keeper of archives. There are no "special effects" in them (what Freud called *dream-work*), no flying through the air, no objects changing before your eyes, no gross distortions that say this is only fantasy. That's why older people are so proud of the age and clarity of a mental picture that goes back 60, 70, or 80 years, a picture they can prove they were not just told about.

When researchers John Kihlstrom and Judith Harakiewicz collected first memories from 314 high-school and college students, they found that most memories fell into the categories of trauma (a childhood accident, for example), transition (such as a move to a new home or the birth of a sibling), and trivia (such as sitting on the beach playing in the sand). But clinicians following in the tradition of neo-Freudian Alfred Adler approach early memories in quite a different way. In Adler's words, a person's first memory is "his subjective starting point, the beginning of the autobiography he has made for himself." Our earliest memory is our most important myth, the one that says, "This is how 'I' began."

A woman of 44 who told me of a lifelong fear of displeasing other people knew exactly what her earliest memory was. As a toddler, she had stood behind a glass door and watched an angry mother walk out on her. "I can still see the sidewalk that she walked down, and the cracks in the sidewalk," she said. "That had a real lasting effect on me. It was like, 'You do what I say or I'll leave you.'" It has taken this woman many years to realize that her life does not have to end in the same mythic place where it began. She doesn't have to be afraid of people walking out if she fails to do their bidding.

The subjective starting points established by our first memories are like the creation stories that humans have always told about the origins of the earth. In some of these stories, the earth developed from a mother who sacrificed herself so that we might live off the nourishment of her body. In others, our world came from the intercourse of Father Sky and Mother Earth, or from a cosmic egg, or from a turtle rising like an island from the sea, or from the Word of a purposeful deity. The myths differ, but they have something in common. They represent people's efforts to say what their identity is, where they belong, and how they ought to live. "This is who we are," the creation stories say, "because this is how we began." In a similar way, the individual self—knowing how its story is coming out—selects its earliest memories to say, "This is who I am because this is how I began." The self says, for example, "I have always been afraid of displeasing others 'because' my mother walked away from me when I was three."

There are other "firsts" that serve the same mythic function as our earliest memories, establishing how themes in our stories and aspects of our selves develop. For instance, research has shown that far more alcoholics than nonalcoholics remember their first drink and remember it as one of life's most significant episodes. The memory of that drink is a "first time" that stands for other times. It's a creation story that underwrites a present identity: "I am an alcoholic."

Most of life's significant firsts (such as your first day of school, your first date, the first time you made love, your first job, the first home of your own—even the look on your mother's face when she first saw your baby) usually occur during the period that will eventually form the hill of reminiscence. These are the years of expansion in a life, the period when we try out new roles, make mistakes, and gradually shape our identities. By the age of 30 most of us have made the choices that count, and by the age of 50 we are beginning to see their long-range consequences. Why does a reminiscence hill appear in the second half of life? Because the maker of myth is turning its attention to our origins. Knowing how our story is coming out, it's setting up a beginning that will explain the ending.

There is variation in all of this, of course, variation in the contents of our memories, variation in the years that we remember best. But the variation speaks to who we are. And there's more to the mystery of memory than the stories of individuals. There's the collective remembering that goes on in families, tribes, and nations—with the same tension between the keeper of archives and the maker of myth. What is memory like when it goes beyond mere *auto*biography—when we tell our children stories of our parents and grandparents? Another great puzzle lies in that question, another great story in the answer.

THE GRANDEUR
OF THE GRAND CANYON

DIANE ACKERMAN

Nothing prepares you for the visual thrill of sailing over the rim, from a state of flatland predictability suddenly into one of limitless depth, change, and color. All at once we are down in its jungles of rock, plunging toward sheer crevices, skimming limestone jags by only a few yards, then swooping down even farther to trace the winding path of the Colorado River, rocketing up toward a large butte, wing left, wing right, as we twist along the unraveling alleyways of rock, part of a spectacle both dainty and massive. Who could measure it, when we are the only object of certain size moving through the mazes? Off one wing tip, a knob of limestone curves into arrowhead edges and disappears at the base of a half-shattered tree, whose open roots catch the sunlight in a cage of iridescence.

When we land, we begin to explore the Grand Canyon on foot from lookouts and trails along the rim. Hypnotized by the intricate vastness, I hike from one triangulation station to another, finding two of the ninety bronze survey disks that were installed fifteen years ago by the Boston Museum of Science–*National Geographic* expedition. Sitting alone on a plinth jutting far out over the emptiness, I listen to the monumental silence and find my mind roaming over the notion of wonder. The canyon is, in part, a touchstone to other wonders, revealing the uncanny work of erosion, a great builder of landscapes. Five geologic eras are here piled one on top of the other like Berber rugs, the evolution of life viewable in a fossil record. Gigantic as the canyon is (217 miles long), it is the world in miniature: seven environments (from Sonoran to Arctic Circle), desert barrenness to spring lushness. It is certainly the grandest American cliché, explored by many but an enigma nonetheless. No response to it seems robust enough.

In a world governed by proportion—in which the eye frames a moment, digests it, frames another—scale is lost: visual scale, mental scale, emotional scale. If your lips form a silent *wow* at the sight of Niagara Falls, what is suitable here, where your heart explores some of its oldest dwellings? How can you explain an emptiness so vast and intricate, an emptiness so rare on this planet? Not the sprawling, flat, oddly clean emptiness of a desert or arctic region, but an emptiness with depth. There are no yardsticks, unless one is lucky enough to catch sight of a dark speck moving along the canyon floor, which is a mule and rider. But that is part of the puzzle of this labyrinth, a maze both of direction and of proportion, a maze in three dimensions.

It's easy to forget how ugly nature often seemed to people before Romanticism reexplored the unevenness of natural beauty. Early-nineteenth-century writers found the canyon grotesque—not just dangerous and obstructive and rife with bloodthirsty Indians, but actually a vision of evil. C. B. Spencer described it as "Horror! Tragedy! Silence! Death! Chaos! . . . a delirium of Nature," while another writer called it "the grave of the world." After two World Wars and assorted smaller ones, with all the atrocities attendant to them, it's no longer possible to find works of nature horrible, tragic, deadly, chaotic; mankind has personalized those traits forever. Now the canyon is just the opposite: a sanctuary, an emblem of serenity, a view of innocence.

The Cárdenas expedition of 1540 discovered the canyon for the Caucasian world but felt no need to name it. For three hundred years it was too overwhelming to report except in whole phrases and sentences. And then in the 1850s and 1860s, "Big Cañon" and "Grand Cañon of the Colorado" came into use, as if it were *one* of anything. For it is not one but thousands of canyons, thousands of gorges and buttes, interflowing, mute, radiant, changing, all with a single river among them, as if joined by a common thought.

In the canyon's long soliloquy of rock, parrots of light move about the grottoes, and real swifts loop and dart, white chevrons on each flank. The silence is broken only by the sounds of air whistling through the gorges and insect or bird song. Now and then one hears the sound of a furnace whumping on: a bird taking flight.

There is no way to catalog the endless dialects and languages and body types of the tourists encountered at the rim. With binoculars as various as they are, visitors search the canyon for trails, mules, signs of other people. The need to humanize the marvel is obsessive, obvious, and universal. With glass lenses extending real eyes, canyon visitors become part of the evolution on show. If we cannot go backward in time, we can at least creep into it, above desert floors and red-rock mesas and ponderosa pine, then suddenly slip over the rim of dreams and down through the layers of geological time.

What is *grandeur* that the word should form rapidly in the mind when one first sees the Grand Canyon? Why do we attach that concept to this spectacle? Is it merely the puniness of human beings compared with the gigantic structures of rock? The moon, the biggest rock most of us know, has been domesticated in literature and song, but the canyon has resisted great literature. Like the universe and the workings of nature, there is no way to summarize it. The ultimate model of a labyrinth, it is gargantuan and cryptic, full of blind alleys and culs-de-sac. We are compulsive architects; to see engineering as complete, colossal, and inimitable as this —still far beyond our abilities—is humbling indeed. As John Muir said in 1896, upon first viewing the canyon, "Man seeks the finest marbles for sculptures; Nature takes cinders, ashes, sediments, and makes all divine in fineness of beauty—turrets, towers, pyramids, battlemented castles, rising in glowing beauty from the depths of this canyon of canyons noiselessly hewn from the smooth mass of the featureless plateau."

Most of all, the canyon is so vastly uninvolved with us, with mercy or pity. Even the criminal mind is more explicable than this—a quiddity we cannot enter, a consciousness that does not include us. We pass through much of our world as voyeurs, and yet we are driven, from sheer loneliness, I suppose, to attribute consciousness to all sorts of nonconscious things—dolls, cars, computers. We still call one another totemic names by way of endearment: we would like to keep the world as animate as it was for our ancestors. But that is difficult when facing a vision as rigidly dead as the Grand Canyon. It is beautiful and instructive and calming, but it cannot be anything other than it is: the absolute, intractable "other" that human beings face from birth to death.

In between consciousness and the Grand Canyon, matter has odd fits and whims: lymph, feathers, Astro Turf, brass. Cactus strikes me as a very odd predicament for matter to get into. But perhaps it is no stranger than the comb of an iris, or the way flowers present their sex organs to the world, or the milky sap that often oozes from inedible plants. There is something about the poignant senselessness of all that rock, wave after wave of blunt, endless rock, that reminds us, as nothing else could so dramatically, what a bit of luck *we* are, what a natural wonder.

At the south rim, brass sighting tubes make arbitrary sense out of the vista. Lay the lensless tube into a slot marked Battleship, and there will be a facsimile in rock. The other sites are mainly temples: Vishnu Temple, Wotan's Throne, Zoroaster Temple, Brahma Temple, Buddha Temple, Tower of Ra, Cheops Pyramid, Osiris Temple, Shiva Temple, Isis Temple, and so on. One of the most dramatic, tall, and precarious buttes is referred to as Snoopy, because, they say, it resembles the cartoon dog lying on his doghouse. All this taming of the spectacle appalls me. Why define a site with another site that is smaller and in some cases trivial? Nothing can compare with the Grand Canyon, and that is part of its marvel and appeal.

It was John Wesley Powell who, in 1860, gave the salient buttes their temple names. Now that the gods who instructed us are remote, we are quite obsessed with temples. We have moved our gods farther and farther away, off the planet, into the Solar System, beyond the Milky Way, beyond the Big Bang. But once upon a time, when time was seasonal, the gods were neighbors who lived just across the valley on a proscribed mountain. Their deeds and desires were tangible; they were intimates.

Today on the Hopi mesas close to the Grand Canyon, in rituals older than memory, men still dress as kachinas—garish, expressionist re-creations of the essences in their world. There is a kachina of meteors, and one of maize, and one of water vapor. In the winter months the kachinas dwell on the twelve-thousand-foot slopes of Humphreys Peak, and in the growing season they come down to move among men. The Hopi have traditionally traveled down into the canyon to perform some of their rituals, and there is a spot on the bank of the Little Colorado where they believe humans may first have entered the world.

Indeed, the whole area around the Grand Canyon is full of lore and natural wonders. The volcanic field just north of Flagstaff is the largest in the United States, and flying over it you can see where the black paws of lava stopped cold. The aerial turbulence at midday evokes the early turbulence from which the canyon was partially formed, and long before that the chaos of the Big Bang. At the Lowell Observatory, in Flagstaff, Pluto was first sighted. In half a dozen other observatories, astronomers cast their gaze upward while, close by, a million tourists cast theirs down into the canyon.

There would be no canyon as we perceive it—subtle, mazy, unre-peating—without the intricate habits of light. For the canyon traps light, rehearses all the ways a thing can be lit: the picadors of light jabbing the horned spray of the Colorado River; light like caramel syrup pouring over the dusky buttes; the light almost fluorescent in the hot green leaves of seedlings. In places the canyon is so steep that sunlight only enters it, briefly, at noon; the rest is darkness.

It is hard to assimilate such a mix of intensities; it is too close to the experience of being alive. Instead, we order it with names that are cozy, trendy, or ancient. Available, viewable, definable, reducible to strata of limestone and fossil, they are still mysterious crevices, still unknowable, still overwhelming, still ample and unearthly, still the earth at its earthiest.

The Douglas fir crop out, under, around, between, through every place one looks; they survive the rock. Many of their twisted, lightning-licked limbs are still in leaf. The cottonwoods, growing over a hundred feet tall, can use more than fifty gallons of water each day. There are a thousand kinds of flower and species of squirrel and bird indigenous to the canyon (some nearly extinct). And endless otters, skunks, beavers, ring-tailed cats, deer, porcupines, shrews, chipmunks, rats, and wild burros. In the low, common desert of the inner canyon depths, only the prickly-pear cactus survives well the high temperatures and rare precipitation. It is not erosion on a large scale that has formed the canyon but small, daily acts of erosion by tiny plants and streams, reminding us what the merest trickle over limestone can achieve. From rim to floor, the canyon reveals the last two billion years of biological and geological history and thus typifies the processes of evolution and decay in which we all take part.

But mainly there is the steep persuasion of something devastatingly fixed, something durable in a world too quick to behold, a world of fast, slippery perceptions where it can sometimes seem that there is nothing to cling to. By contrast the canyon is solid and forever; going nowhere, it will wait for you to formulate your thoughts. The part of us that yearns for the supernaturalism we sprang from yearns for this august view of nature.

At nightfall, when we reboard the plane for our flight back to Phoenix, there is no canyon anywhere, just starry blackness above and moorish blackness below. Like a hallucination, the canyon has vanished, completely hidden now by the absence of light. Hidden, as it was from human eyes for millennia, it makes you wonder what other secrets lie in the shade of our perception. Bobbing through the usual turbulence over the desert, we pick our way home from one cluster of town lights to another, aware from this height of the patterns of human habitation. Seven skirts of light around a mountain reveal how people settled in waves. Some roads curve to avoid, others to arrive. Except for the lights running parallel along the ridges, people seem desperate to clump and bunch, swarming all over each other in towns while most of the land lies empty. The thick, dark rush of the desert below, in which there is not one human light for miles, drugs me. Looking up drowsily after a spell, I'm startled to see the horizon glittering like Oz: Phoenix and its suburbs. That, or one long, sprawling marquee.

SAVE THE SWAMPS!

LET THE WETLANDS FEND FOR THEMSELVES

JAMES GORMAN

I'm a fan of swamps. Not wetlands. Swamps. I like them dark and humid, with tea-colored water and snakes. Also alligators and escaped convicts. My wife and I spent part of our honeymoon in a swamp. We left a resort filled with southern fullbacks who had just married cheerleaders and spent a day cruising the Okefenokee. We got soaked in a massive downpour and sat steaming in the boat afterward, staring at a spectacular rainbow. We saw alligators and turtles and birds and Spanish moss. At one point we were in a perfectly quiet canal when all at once an alligator slid off a bank, a flock of ibises took off from the trees, and a hawk passed overhead. Not quite the high point of the honeymoon, but a good moment nonetheless.

Since then we have traveled *en famille* to the Everglades more than once, and wherever we are we always find a bog walk or marsh slog to take. In the Everglades I once spent fifteen minutes watching a great blue heron try to eat a snake. At the time I left, the snake was still alive, wrapped around the heron's beak. I would have liked to have stayed for the conclusion, but I had appointments to keep with anhingas and purple gallinules. If I had it my way, we would vacation every year in or near a swamp. So I figure since somebody has to speak up for swamps and liberate them from the deadly wholesome aura of wetlands, it might as well be me.

I trace my love for swamps—for that strange smell that marsh muck has, for the sound when your foot comes out of it, for the sense that every inch of water and vegetation is literally crawling with something—to my boyhood. There were a number of little bogs and marshes where I fished or speared frogs or just slogged around. But my first and best swamp was a marshy bit of Connecticut woods with a seasonal pond that was a hideout for me and my friends. Wet land. I suppose certain people would describe it as some sort of palustrine wetland. But for us it was the Swamp, and it grew tadpoles in biblical numbers and frogs uncountable, including one that is indelibly recorded in my memory.

This particular frog was in a small backwater of the swamp, surrounded by smaller frogs as if it were some malevolent amphibian goddess. In memory it is the size of a round loaf of peasant bread, of a basketball, of a desktop computer. I only saw it once and I remember being afraid to try to kill it, afraid even to get into the water with it. I have no idea how big the frog really was. The memory, as vivid as it is, is all mixed up now, due to the passage of time and twisting of neurons, with one of Hans Christian Andersen's scariest fairy tales—"The Girl Who Trod on a Loaf." This tale is about a girl named Inger who, in order to save her clothes from a bit of marsh muck she has to walk over, puts down a loaf of bread she is bringing home to her poor parents and steps on it. Immediately, she sinks into the Marsh Wife's brewery.

> A scavenger's cart is sweet compared to the Marsh Wife's brewery. The smell from the barrels is enough to make people faint, and the barrels are so close together that no one can pass between them, but wherever there is a little chink it is filled up with noisome toads and slimy snakes. Little Inger fell among all this horrid filth.

I realize that there is no specific mention of a frog. Nonetheless, in my mind's eye I see my amphibian behemoth as gatekeeper and guardian of that noisome brewery and the other horrors that lurk under the surface in swamps, horrors such as Andersen's Bogeyland, where Inger was turned into a conscious but immobile statue:

Her clothes were covered with slime. A snake had got among her hair and hung dangling down her back. A toad looked out of every fold of her dress, croaking like an asthmatic pug dog. It was most unpleasant.

Don't misunderstand. As gruesome as these passages are, I'm quoting them as evidence not of the awfulness of swamps but of their mysterious appeal. Our swamp had no Marsh Wife beneath the surface (as far as I know), but it had snapping turtles of a size to snap a broomstick in their jaws, and that was part of its allure; it was just scary enough. In the spring and early summer my friends and I would ignore parental cautions and cross the railroad tracks behind my house. Then we would follow well-worn paths through the woods to reach the swamp. Sometimes we would spear frogs; sometimes we would catch them by hand. At other times we would catch turtles, although not the snapping kind. There was an old raft that we poled around the swamp, trying to push each other over in hopes that somebody would either get sucked into quicksand or lose his big toe to a snapping turtle.

One day, on the train tracks near the swamp, a friend and I found a big snapping turtle dead and disemboweled. I have no doubts about my memory in this case. I know that this turtle was as big around as a seventeen-inch television screen. We assumed its death was the work of the older boys who ventured into the woods with BB guns and .22-caliber rifles instead of the pocket knives and frog spears that my friends and I carried. The turtle was the first evidence of the gruesome death of anything other than a small amphibian or a large insect that we had yet seen. It disturbed us, but I have to say it also impressed us with the prowess of the disembowelers, who had, after all, somehow captured this creature and carried it to the tracks where they had smashed it up, all presumably without injury to themselves.

Deep in thought, examining the turtle's innards, we failed to hear the train coming until it was very close. It was the same slow freight that always came by, but this time it seemed faster than usual. All of a sudden the engineer pulled his horn and the engine was right upon us. We ran in front of it, dodging back and forth, trying to decide which side of the berm we would jump on, laughing hysterically, with the horn blowing more frantically until we both jumped on opposite sides. I lay there watching the freight's wheels churn by only a couple of yards from my face.

This was one of my formative childhood experiences. Of course, it occurred near the swamp, not in it, but it has always seemed swamp related to me. Death, danger, innards, freight trains, swamps—I feel they're all connected in some as yet unrealized ballad by Merle Haggard. The song of the swamp. And this is the great thing about swamps in my mind—that they feed the imagination and are fed by it. My swamp was a combination of concrete ecological circumstance and the swamp lore and literature that, even as a child, I had absorbed. A swamp—unlike a wetland—is not an objective, measurable, scientific entity, but a malleable mixture of place and legend.

Swamps enrich the language. They serve as a source of metaphor. Are all of us, black and white, caught in a swamp of soured race relations? Yes. Do we feel wetlanded by these problems? Well, not exactly. There is swamp oak, swamp cabbage, swamp mallow (a different genus and species from the marshmallow, by the way), swamp fever, and a thousand other swamp things, not to mention Swamp Thing itself, or the famed Revolutionary War guerrilla hero General Francis Marion of South Carolina, the Swamp Fox.

There are even swamp movies, such as the unforgettable *Wind Across the Everglades* and *The Defiant Ones,* to mention two neglected classics. *Wind Across the Everglades* is about feather pirates whose king is Burl Ives. In my favorite scene, Ives, who has suffered some severe disappointment (I think he may have been shot) and knows that his way of life (killing egrets for money, if memory serves) is disappearing, picks up a fairly thick snake and holds it to his arm, saying, "Bite deep, Brother Cottonmouth, bite deep." At least that's what I remember him saying; *Wind Across the Everglades* is not out on video, so I suppose my Burl Ives memories may be as flawed as my frog memories. In *The Defiant Ones,* Tony Curtis and Sidney Poitier, escaped from prison and manacled together, slog through a classic southern swamp scene, sinking waist deep in racial tension. In a much later, elegiac bit of cinematic bog business, there is Jim Jarmusch's nostalgic takeoff of such movie swamp chases in the blackly hilarious *Down by Law*. Three escaped convicts trudge, muck, and pole a leaky boat in circles in some godforsaken bayou, getting angrier and angrier at each other.

By comparison, the lore and language of wetlands are impoverished, bereft of connotation, shackled to denotative science and law. They involve flood plains, environmental-impact statements, vital ecological roles, biota, and water levels. Not that *wetland* is a useless term. It's just, if you'll excuse the paradox, dry. It's an organizing label, a taxonomic tool. There are riverine and lacustrine wetlands, and palustrine ones as well, the deep nature of all of them obscured by the Latinate language of science.

What may serve a researcher well serves the soul poorly, and the rhetorical warfare of the environmental cause not at all. I'm sure that part of the popularity of *wetland* has derived from the notion that a *swamp* sounds bad and steamy, not the sort of place civilized farmers and city folk want to venture into. More the sort of miasmic hotbed of disease and swamp gas that was made to be drained in the same way that the *jungle* (now the cleaner, gentler *rain forest*) was made to be razed. A wetland; there's something that sounds worth saving. Well, maybe. But I don't think the new labels have worked. It seems to me that the rain forest is disappearing pretty fast, as fast as when people mistakenly called it jungle. Wetlands likewise.

The terms *rain forest* and *wetland* just don't have sex appeal. These terms appeal to the ego and the superego. The pleasure offered by contemplation of a wetland and its place in the web of nature is cerebral. The satisfaction of trying to preserve it is the even, smug glow that comes of doing good. Not a glow in great demand these days. Ah, but swamps, what do they appeal to? Something dark and humid, I'd say, something arising from the reptile brain itself (obviously a fan of swamps), something that likes to watch music videos on MTV.

In fact I would go so far as to say that swamps not only appeal to the id but are its embodiment in the natural world. If we were to parse landscape in Freudian terms, then the ego would be sunlit meadows; the superego, craggy peaks to be savored only after punishing climbs; and the id, a powerfully aromatic swamp—fecund, moist, mysterious, crawling with phallic serpents and alligators, warm and dark, inviting penetration but always threatening to swallow one up, to eat one alive.

No doubt a few healthy-minded hikers and the great majority of graduate students in the life sciences will be inspired to save a wetland or two by talk about ecological balance and migrating waterfowl. But not only good people are needed to save swamps. A movement restricted to good people cannot hope to survive. Swamps, unlike wetlands, can play to a mass audience. They have what the public wants—sex, violence, snapping turtles. I happen to believe that advertisements that emphasize how dangerous drugs are only make them more interesting to teenagers. Teenagers like to think they are flirting with death. Well, let them flirt with swamps. Let's not talk about ecologically valuable wetlands, let's talk about murky, dark, snake-ridden soughs and sloughs, bogs and bayous. Think about it: goody-goody heroes—and landscapes—are admired by teachers, parents, and public-interest groups. But everybody loves an outlaw.

Save the Swamps!

THE GREEN FUSE

CHRISTOPHER JOYCE

The photograph staring up from the crumbling page of an old botany textbook is arresting. Ghostly and tall next to the Indian, the white man stands shirtless, his hair close cropped and his chin stubbled, his biceps large against his spare frame. With utter concentration, he is inserting one end of an eight-inch bone tube into his right nostril. The other end of the tube is held between the lips of the Yukuna Indian, a compact man with an air of knowing exactly what he is doing. Inside the tube is a reddish brown powder, a powerful and toxic narcotic made from the bark of an Amazonian tree. The white man rests his right hand on the shoulder of his sturdy instructor, perhaps to steady himself, while his partner gently grips the snuffing tube between thumb and forefinger. The Indian takes a deep breath . . .

The Yukuna calls the snuff *a-re-dje,* although in his neighborhood, the Colombian Amazon, it has many names, like *yákee* or *yáto.* The white man knows it as *Virola,* in this case from the tree *Virola calophylla.* The Yukuna employ it to prepare for the *Kai-ya-ree* dance, a three-day ceremonial reenactment of the evolution of their tribe, an evolution that began, of course, with the primordial anaconda. Like most magical Amazonian plants, *a-re-dje* has an animate spirit and puts the user in touch with the divinities. The white man, however, believes it contains about a half-dozen indole alkaloids belonging to the open-chained or closed-ring tryptamine derivatives with a tetrahydro-B-carboline system.

After all, he's a Harvard man.

He is, in fact, Richard Evans Schultes, the century's preeminent ethnobotanist, or scholar of the use of plants by native peoples. He spent thirteen years during the 1940s and 1950s living with Indians in the Amazon, mostly in Colombia. During that time, he observed lots of medicine men taking *Virola* and similar "plants of the gods." He wrote in his first scientific paper on *Virola*:

> The dose employed by the medicine-men is sufficient to put them into a deep but disturbed sleep, during which delirious mumblings or, sometimes, shouts are emitted; visual hallucinations or dreams are reported to accompany the narcotic sleep very often. These are "interpreted" by an assistant who awaits the prophetic or divinatory sounds. Some medicine-men, it is said, are affected more violently than others and uncontrollable twitching of the fingers and facial muscles and a popping of the eyes are not infrequent symptoms. There is one report of [a] death. . . .

Schultes tried one-quarter the normal dose. He got a headache, his feet and hands got numb, and he felt like throwing up. He was overcome with lassitude and uneasiness, and he fell into fitful sleep accompanied by heavy sweating, especially in the armpits. After twenty hours of this, he concluded that the snuff had its "shortcomings." It was, without doubt, a hallucinogen and a dangerous narcotic.

Schultes ingested quite a lot of strange things while working in the Amazon. Food, of course. "It was monotonous, but good, actually," he recalls, with game occasionally adding variety to staples like boiled tubers. One could always wash it down with *chicha,* an alcoholic drink fermented with saliva, if one were so disposed. As for experimental substances, besides *yákee* there was the quite awesome *ayahuasca,* the "vine of the soul" that liberates the user from everyday life (itself a hallucination) and introduces him to what he will consider reality, a place of wonder and sensuousness where he can converse with his ancestors and learn the meaning of his world from the jaguar, the snake, the monkey, and the vine. "It is a form of communion," explains Schultes, who, of course, tried it.

Schultes is now seventy-seven years of age. Twice a week, he climbs four flights of stairs to his office at Harvard's botanical museum, where he is curator and professor emeritus. He still wears a white lab coat over his Harris tweeds and khakis, even though he spends most of his time writing letters, a chore that comes along with the accolades given those scientists who reach the pinnacles of their discipline. He'd just as soon go back to Colombia. "But my wife won't let me," he says ruefully, if not quite seriously. "She reads in the newspapers that people are being shot in the streets down there." As the eminence of ethnobotany, he has been a man of two centuries, recasting the science for the twentieth yet practicing it like the naturalist explorer of the nineteenth. "He really is a Victorian," says his colleague and longtime friend Robert Raffauf, a chemist who's spent many an hour himself drifting down the Amazon in a dugout canoe.

Indeed, in the reminiscences Schultes is asked to share with scholarly convocations or visiting journalists, he usually invokes the name of his spiritual mentor, Richard Spruce. Spruce was a Yorkshireman who in 1849 set out for the Amazon to collect plants for Kew Gardens and a few wealthy collectors who paid his way. Spruce spent seventeen years trekking and gathering. The journey earned him little money and only minor fame outside botanical circles, and it very nearly killed him on several occasions. "He was a true naturalist explorer," says Schultes, whose father read to him as a child from Spruce's biography.

While Schultes did retrace many of the footsteps of the nineteenth century's botanical pioneers, he also broke new ground. Not satisfied with simply collecting and classifying, he learned how the plants were put to use. The western Amazon was a hothouse of cultures: the Tukanos, Gwananos, Guahibos, Taiwanos, Kubeos, Boras, Andokes, Witotos, Yukunas, Tanimukas, Matapies, Puinaves—the diversity seemed as rich as that of the flora. Each tribe possesses an unwritten encyclopedia of what we call *economic botany,* the practical use of plants. To the Indians, the flora are quite a bit more, however. Schultes writes in his book *Plants of the Gods* of an Indian story of *yopo,* an hallucinogenic snuff:

> In the beginning, the Sun created various beings to serve as intermediaries between Him and earth. He created hallucinogenic snuff powder so that man could contact supernatural beings. The Sun had kept this powder in his navel, but the Daughter of the Sun found it. Thus it became available to man—a vegetal product acquired directly from the gods.

The flora and fauna of the Amazonian forests fed, clothed, and housed hundreds of cultures for millenia. The very tissues of the plants came from the gods and provided medicines and, like *yákee* and *ayahuasca* and the dozens of other hallucinogens, the means to interpret messages from both god and forest. The essentials of myth and religion arose from root and bark, flesh and feather, puddle and torrent.

If the Earth were man-made—and we who call ourselves civilized seem to be headed in that direction—then these forests and their cultures would be baroque cathedrals. Nowhere on the planet lives such biological intricacy and variety, what biologists now call *biodiversity.* Perhaps nowhere do people live so closely and harmoniously with their natural surroundings. And within our lifetimes, these forests will all be gone.

Many people will take as little notice of the loss of biodiversity as they do of the drone of traffic outside their windows. Others will enjoy more wealth from cattle ranches and farms. Others will consider it the sad but inevitable price of progress. But for most it won't change the meaning of life.

Or will it?

There is one who believes it will. To find him, one need go no farther than a few steps from Schultes's Harvard office. Down one floor, past the cases of intricately crafted glass flowers and into the museum of comparative zoology, past dioramas of skeletal saber-toothed tigers and stuffed black bears frozen in midgrowl, across a closed catwalk and into the next building, until you reach the realm of the insect. Specifically, the ant.

Ant master Edward O. Wilson resides here. It was here in his infested office—the ants outnumber him by several thousand to one, crawling through an ant city he has constructed out of plastic boxes on a row of shelves—that he changed the way people think about genetics. His book *Sociobiology,* published in 1975, proposed that human behavior can be influenced by our genes. Some people didn't like that; it smacked of genetic determinism, eugenics, biological destiny. They told him to go back to his ants. Wilson, a soft-spoken Southerner who is polite to his undergraduate students and who insists that all he really wants to do is study his ants, kept hammering away, publishing several books to clarify his views. He didn't convince all the nay sayers, although he did win a Pulitzer Prize for his book *On Human Nature.*

In that book and others that followed, Wilson talked about certain behaviors that appear to be universal among organisms, for example, selecting habitats to raise one's chances of survival. Habitat selection applies to both the instinctive journey a peanut-sized newborn kangaroo makes from its mother's vaginal opening to her nipple, located deep in her pouch, and the automatic preference of the alder flycatcher for swamps and wet thickets. "If you get to the right place," says Wilson, "everything else is likely to be easier."

Do humans possess such an innate homing instinct? Wilson thinks we might. He would classify it as part of his *biophilia hypothesis.*

"The biophilia hypothesis is that human beings have a deep and complex innate propensity to enjoy and affiliate with a wide diversity of life forms. And specifically, life forms . . . in the midst of which the human species has evolved over millions of years," says Wilson. The archaeological evidence is clear as to what kind of terrain we preferred when we came down from the trees and learned to love the feel of the Miocene mud beneath our feet: the tropical savanna, a landscape with bodies of water, rolling plains of grass, stands of trees for shelter and food, and a few hills or escarpments that provided the odd cave or two and a place from which to watch for predators and prey. It was the kind of place in which a hunter-gatherer could thrive.

The body, then, evolved within and became predisposed to this kind of habitat. But, asks Wilson, is the *mind* also partial to the tropical savannah? Is there some canvas of biotic beauty that, having served as the scrim against which we played out the human drama for millions of years, can be said now to "lie in the genes of the beholder?" Like our apparently innate fear of snakes, it's an observation we make but cannot prove convincingly. Wilson is, however, quite happy to speculate. "It seems that whenever people are given a free choice, they move to open, tree-studded land on prominences overlooking water. This worldwide tendency is no longer dictated by the hard necessities of hunter-gatherer life. It has become largely aesthetic, a spur to art and landscaping," he writes in his book *Biophilia*. Japanese teahouses, monuments to miniaturization, artfully landscape closet-sized gardens behind windows to create a *trompe l'œil* of depth and space. Gardens graced almost every inn, restaurant, and residence in Rome's Pompeii. Most had the same layout: trees and shrubs, beds of flowers, pools and fountains. Monuments from the Mall in Washington to the palaces and formal gardens of Europe and Asia re-create the same sort of savannah gestalt.

We seem to yearn for this contact with our ideal of the natural world no matter how far removed we become from it, and we will go to great lengths to satisfy the urge. Try looking out from atop a Manhattan skyscraper and spotting the islands of green topiary gracing penthouse retreats thirty stories up in the sky.

So why dilate on what seems obvious? The question brings us back to biodiversity. We need a green environment to thrive, and not just any green. We need biological diversity, not just because it's pretty or because without it a few biologists in muddy boots would lose their favorite hunting ground, but because we as a species are part of it and it is part of us. Extinguishing biodiversity may carry some unforeseen consequences beyond simply losing a potential cure for cancer or a novel new fruit. Says Wilson, "We may be tampering with a part of the environment that future generations will discover to have been very important for the full development of the human psyche as a spirit."

The Indians of the Colombian Amazon certainly don't need to be told that their spiritual well-being is umbilically entwined with the well-being of the forest. Nor, for that matter, do people like Edward Wilson, who gladly admits to holding an almost religious connection with nature whenever he goes out into the field. He calls it the naturalist's trance, a sense of heightened awareness and expectancy. He experiences it simply walking into a rain forest, although he becomes especially entranced, he adds, when crawling around on his hands and knees in the leaf litter, exploring the microwilderness for bugs. That transcendent experience is cut from the same cloth as that of the Witoto on the hunt for peccary or the Yukuna in his hallucinogenic communion with his makers. And so too is our own yearning, albeit ever more repressed, for that which is loamy, green, earthen, and fecund.

Richard Schultes has traveled in the world of our living ancestors and reported back. His teaching laboratory at Harvard is a shrine to their lore and the bounty of the forests: the dried leaves of *Banisteriopsis,* source of *ayahuasca;* the blowgun darts and the bark of *Curarea* with which they are poisoned; and the snuff tubes that Schultes learned to use forty years ago. "The Indians have knowledge of the forests built on thousands of years of experience, and they still use it," says Schultes. "You must remember when you go into the forest that they are the professors and you are the student."

PULLING THE HANDLE OFF THE PUMP:

SUNLIGHT, CANCER, AND THE BROTHERS GARLAND

PETER RADETSKY

It was a godawful July day in Baltimore. Hot, muggy, gray—nothing like the bright skies back home in San Diego. Besides that, the air-conditioning was hardly working. The laboring fans served mainly to rattle the window shades, noisily interrupting the presentation. And there they were, the wayfaring brothers Garland, cooped up inside the darkened classroom, attending a seminar on cancer.

They had come to Baltimore three years before, in 1974, driving across country to the mecca of their profession, the Johns Hopkins School of Public Health, Cedric for a job as an assistant professor of epidemiology and Frank, his little brother, to enroll as a graduate student in the same department. Chronicling the patterns of disease was their passion.

So when the maps flashed onto the screen, they were immediately intrigued. These were the first maps to show death rates from cancer for all 3,056 counties in the United States. Map after map, cancer after cancer—lung cancer, pancreatic cancer, stomach cancer—with deaths scattered all over the U.S. like buckshot. Epidemiologists look for order, for useful trends, but there was no order to this. Cancer, it seemed, was a random killer.

Then another map flashed onto the screen, and the Garlands, sitting there in the front row, sat straighter. Colon Cancer Deaths, the slide was labeled. In contrast to the maps that had come before, this one showed a definite pattern: below 37 degrees latitude, from the Mason-Dixon line all the way west to San Francisco, the disease hardly existed. It showed up in the Northeast, and heavily; it showed up in the Midwest and northern Great Plains; but in much of the South and West it was almost unknown.

That was strange. What could account for such a discrepancy? No one knew. Then another map appeared on the screen. This one was of skin cancer, and it showed precisely the same pattern, only in reverse. Below the 37th parallel, all sorts of skin cancer—above it, very little. That, at least, made some sense. It was known that sunlight could cause skin cancer, and more sunlight fell in the southern part of the U.S. than in the rest of the country.

It was then that the Garlands experienced their epiphany. Cedric leaned over to Frank and whispered, "I wonder if sunlight might prevent colon cancer?"

Today, back home in San Diego, sitting in Frank's office, which perches on a sandy spit of land covered by sunflowers, creosote, and berry bushes sloping away to the blue Pacific Ocean, the Garland brothers look back on the moment. "I can remember it like it was yesterday," says Frank. "It's what kicked us off on these last fourteen years."

Cedric, 44, now heads the epidemiology program at the University of California at San Diego Cancer Center; Frank, 40, runs the epidemiology program at the Naval Health Research Center outside of town. They are the only brother epidemiologists in the cancer field, maybe the only brother epidemiologists anywhere. They work together, live together, and are fixed on the same goal: proving the insight hatched on that hot Baltimore day.

Says Frank, "If you tell me how much sunlight there is where you live, without knowing anything about your reproductive history, or how much fat or fiber you eat, or any of those things, I can tell you almost exactly what your risk of colon cancer is."

Frank likes to make statements like that. The more flamboyant of the brothers, with the tanned face and sun-streaked hair of a southern California surfer (which, surprisingly, he is not), he is the salesman of the duo. It was a phone call from Frank that first made me aware of the brothers' story; it is Frank who sends me articles and calls me with updates. Cedric, larger, plumper, grayer, and more genteel (he drives a four-door Buick), is the idea man. But don't let appearances fool you. Intellectually he is adventurous, even fearless. "Sunlight prevents colon cancer," he says unequivocally. "We can eradicate the disease."

It is a startling notion. Colon cancer is the second leading cause of cancer death in the United States, striking 150,000 people each year and killing 60,000. In the last decade, scientists have shown that the disease comes about when key genes mutate and cause cells to proliferate uncontrollably, eventually to pile up in the chaotic masses known as tumors. These mutations may be due to a number of influences, including viral infections, exposure to carcinogens, a diet high in fat and low in fiber, even heredity. Lack of sunlight has never been considered one of them —nor has a strategy yet been found to prevent the disease on a large scale. So while the Garlands themselves harbor no doubts about the connection between sunlight and colon cancer, persuading the rest of the scientific community hasn't been easy. They've had to become epidemiological detectives, gumshoeing around the world and into the mysteries of medical history. For it's one thing to come up with a hunch, no matter how inspired—it's another to pin it down.

Soon after the seminar, the Garlands began by digging into that staple of epidemiologists everywhere, statistics. During their journeys home to California they were struck by the difference in amount and intensity of sunlight along the way. "It was the increasing clarity of the atmosphere as we moved from east to west," Cedric explains, "with the gray skies of the East becoming the brilliant deep blue skies of the Southwest." Measurements provided by the Smithsonian Institution, which had been monitoring sunlight levels since the turn of the century, and the National Weather Service reinforced their impressions. Southwestern states such as New Mexico and Arizona receive almost twice as much sunlight as northeastern states like New York, New Hampshire, and Vermont. When the Garlands turned to National Cancer Institute studies to track down the number of colon cancer cases in these states, what they found fortified their suspicions that sunlight prevented the disease: in comparable numbers of people, colon cancer showed up almost twice as often in the relatively gloomy Northeast as it did in sunbathed New Mexico and Arizona.

With that information as a clue, the brothers proceeded to make a map ("Epidemiologists love maps," laughs Frank) illustrating levels of sunlight falling on the country as a whole. They discovered that the pattern was similar to that of the colon-cancer map—you could almost superimpose one atop the other. The Mason–Dixon line of sunlight was 350 calories' worth of mean daily solar radiation (calories, a unit of energy, are commonly used to measure solar radiation). In areas with less than 350 calories, primarily the Northeast and Northwest, there was lots of colon cancer; in areas with more than 350 calories, primarily the South and West, the incidence was way down.

So the connection was valid after all—the more sunlight, the less colon cancer. Maybe sunlight did indeed prevent the disease. "We felt very excited about it," recalls Cedric. "But we had to repress our excitement, because it seemed naive to suggest such a simple solution to what was considered an almost insoluble problem." Besides, the Garlands' findings only pertained to the U.S. If their emerging theory were to hold up, they'd have to test it against data from all over the world.

So again they looked for statistics, this time for worldwide levels of sunlight. "At the Scripps Institute of Oceanography, we found a map in a German book called *The International Atlas of Meteorology*," says Cedric. "It was this very dramatic color map of different levels of sunlight intensities. We could see, for example, the brilliant burning sun of the Sudan and the southwest United States, and the cooler hues of northern America and northern Europe." As the colors on the map faded from an intense orange-red near the equator to a pale yellow in countries more distant from it, they indicated the varying levels of sunlight throughout the world.

Information about the worldwide frequency of colon cancer was

harder to come by. The World Health Organization provided some figures, but the brothers had to dig up the rest for themselves on side trips
during international conferences and other jaunts. "I remember going to
Marrakech and visiting their health department, which was located in a
barracks on the sandy border of the town," says Cedric. "It turned out
that the health officer had been trained at Hopkins. The records were all
in French, maintained in old ledgers, and he let us take them back to our
hotel. There we were, sitting on the balcony of the Hôtel de Paris, this
ramshackle building on the edge of the square, drinking Schweppes Indian
Tonic, and searching these old ledgers for cancer deaths."

What they found during such forays excited them even more. "Every
country with more than three hundred and fifty calories of solar radiation
striking the ground had comparatively little colon cancer," says Cedric.
"It was true north and south of the equator." For example, sun-drenched
Northern Hemisphere countries as varied as Greece, Nigeria, Costa Rica,
and Morocco reported infrequent colon cancer; so did sunny subequa-
torial African and South American countries and the more northerly parts
of Australia. On the other hand, New Zealand, way down around 40
degrees south latitude and with far less than 350 calories of sunlight, had
very high rates of colon cancer, just as the cloudy, rainy United Kingdom
up north had a lot of the disease. And so it went, country after country
—until they came to Japan.

There was Japan—mountainous, rainy, ofttimes cloudy and gray,
much of the country above the 37th parallel, receiving no more, and
sometimes less, than 350 calories of sunlight per day. It should have been
a risky place for colon cancer, right? Wrong. "It has an extremely low
risk," Cedric says, shaking his head. "It's had an extremely low risk for
a long time."

Peering out at the broad Pacific, as though their gaze might take in
Japan itself, the Garlands laugh at the memory. What I want to know is
why? After all, Japan had thrown a monkey wrench in the machinery.
What is there to laugh about?

Not to worry, explains Cedric. "There are exceptions to any theory, and we were very interested in the exceptions." Frank nods vigorously. "You get an overall pattern, then look for unusual incidents. That's down-home epidemiology. That's how John Snow stopped cholera in London."

It's a famous story in medical circles—the Garlands revere it as a model of how to conduct an epidemiological investigation. In the middle of the nineteenth century, a devastating epidemic of cholera descended upon England. Snow, a London anesthesiologist who lived in the Soho district of the city, took it upon himself to look into a local outbreak that had killed over 500 people in just ten days. The prevailing theory at the time was that cholera was caused by miasmas, foul emanations wafting through the air. Snow suspected otherwise. He noticed that in the heart of the afflicted neighborhood, at the intersection of Broad and Cambridge streets, was a well renowned for its water's taste and supposed purity. In those days when most houses lacked piped water, the Broad Street pump was used by virtually everyone in the neighborhood. Rather than these vague miasmas, might not the disease be the result of something in the water? Snow decided to find out.

Snow drew a map of the area, marking deaths with tiny coffins. He found that in the streets surrounding the pump they clustered like iron filings around a magnet. The circumstantial evidence was compelling. But then Snow was confronted with exceptions. Among them was the mystery posed by the 70 employees of a nearby Broad Street brewery—none of them had come down with the disease. If the water were at fault, surely some of them should be dead.

Snow interviewed the brewmaster, and soon the solution to the puzzle became clear: *The men drank only beer.* As the brewery had its own well, they had never once used the Broad Street pump. That was why, in the midst of death, they had survived. But what about the death from cholera of an old woman, a former resident of Broad Street, who had moved miles away to Hampstead some time before? There was no cholera in her new neighborhood—why had this woman been stricken? Snow took a carriage ride to her home and asked her son if there had been anything unusual about his mother's habits. Nothing, the son replied, except for one minor idiosyncrasy. It seemed that from her earlier residence in Soho the old woman had developed a taste for water from the Broad Street pump. To keep her in good spirits, the family had paid a porter to fetch a large bottle of pump water every day. So although the woman did not live in the area at the time of her death, in effect her coffin belonged there. She, too, had been drinking the water.

The exceptions proved the rule. Snow convinced city officials to remove the handle from the Broad Street pump, and the epidemic subsided.

Like John Snow, then, the Garlands set out to investigate their own exception. Why were there so few cases of colon cancer in Japan? One by one they eliminated the possibilities. Faulty reporting of cancer cases? No. More hours of sunlight than normal? No. Unusual concentrations of cosmic rays? No. Genetic resistance? No. "We knew that the reason wasn't genetic," says Cedric. "Japanese immigrants to the U.S. take on rates that are as high as those of people born here. So it had to be something that was present in Japan and not elsewhere."

The answer came when the brothers decided to look at what Japanese people eat. "We found," says Cedric, "that about ninety percent of their protein intake was from fish."

"And," adds Frank, "the fish the Japanese eat is often very high in vitamin D. They take in huge amounts of vitamin D."

Vitamin D. It comes to us from two sources. Either you eat it or *you get it from the sun.* It's created in the skin in the presence of ultraviolet B light, which comes from the part of the light spectrum that also causes sunburn. Lacking ample sunlight, the Japanese were getting vitamin D from food: mackerel, herring, sardines, eel, tuna, and a variety of small dried fish consumed as snacks—all of which are high in vitamin D. Perhaps vitamin D, not sunlight per se, was the common denominator.

Now the brothers were really in deep. To propose that sunlight somehow prevented colon cancer was dubious enough, but to bring vitamin D into the mix . . . That was tantamount to saying that cancer, this dread disease, was instigated by a vitamin deficiency. Certainly there were vitamin-deficiency diseases . . . but cancer? In this age of vitamin supplementation? "Everybody thought we were crazy," says Cedric.

Of course, everybody had thought E. B. McCollum was crazy also. McCollum was a Hopkins researcher, long since dead, whose former office the brothers occupied. He was something of a legend on campus for his groundbreaking work with perhaps the most famous vitamin-D-deficiency disease of all—rickets.

Frank loves to talk about rickets. In fact, it didn't take much to persuade him to give me a slide show about rickets, graphically illustrating the coal-smoke-fouled air of seventeenth-century London, which suffered particularly massive onslaughts of the disease; various historical theories as to its cause; and, once again, the surprising entry of Japan into the picture. When an English missionary named Theobold Palm visited Japan in the nineteenth century, a century in which rickets was so prevalent that half of all children living in English cities suffered its bone deformities, he discovered that while, as in England, there was filth, poverty, overcrowding, and a great deal of disease, Japan had no rickets.

"About this time, eighteen ninety," says Frank, "people in Europe began to use cod-liver oil to prevent rickets. It was an old-wives'-tale remedy that worked. Two decades later, E. B. McCollum set out to determine what it was in cod-liver oil that stopped the disease. That's how he discovered vitamin D." To test his finding, McCollum fed some laboratory rats a diet that included vitamin D and others a diet lacking it. Sure enough, the deprived rats came down with rickets. He also demonstrated the importance of vitamin D produced by sunlight, as he carried some rats to the roof of the School of Public Health for an hour or two a day while leaving other rats in the lab. The outdoor rats' ears would peel from sunburn, but they didn't get rickets. The lab-bound rats, meanwhile, suffered the disease.

Galvanized by his discovery, McCollum visited state after state to urge that milk be exposed to ultraviolet light, thereby becoming fortified with vitamin D. Today all milk in this country is routinely fortified with vitamin D, as well as with other vitamins and minerals. That fact, and the routine administration of vitamin supplements and cod-liver oil elsewhere in the world, has made rickets largely a disease of the past. In effect, McCollum pulled the handle off the pump.

Frank ends his slide show with a flourish. " 'In nineteen eighteen,' " he reads from the screen, " 'rickets was common in England, northern Europe, and the northeastern United States, and was rare in southern Italy, Greece, Turkey, Africa, and, all-importantly, Japan. It was thought to be due to genetics, overeating, a lack of exercise, and infection by microorganisms.' "

The last slide flashes onto the screen. " 'In nineteen seventy-eight,' " reads Frank, " 'colon cancer was common in England, northern Europe, and the northeastern United States, and was rare in southern Italy, Greece, Turkey, Africa, and, all-importantly, Japan. It was thought to be due to genetics, overeating, lack of exercise, infection.' " A look of triumph lights his face. " 'The hidden denominator? Vitamin D.' "

But a geographical association and a historical parallel, no matter how strong, do not make a case. Neither do infectious enthusiasm and over-whelming commitment. When in 1980 they returned to San Diego to take up their new jobs, the Garlands found themselves stalled. They had all sorts of circumstantial evidence that vitamin D, whether generated by sunlight or diet, may prevent colon cancer, but they had no proof. The only way to establish the notion one way or another was to conduct a clinical trial, a so-called intervention experiment, in which they would give vitamin D to some people, wouldn't give it to others, and would keep track of who came down with colon cancer. That would show, once and for all, if vitamin D forestalled the disease. If the results were positive, the brothers could then recommend the vitamin's use as a preventive. But the Garlands had no experience running clinical trials, and before other people would consider planning such an effort they would have to be convinced by more than just a compelling hypothesis, a historical analogy, and a geographical connection. They would need direct evidence, something the Garlands didn't have.

To the rescue came the head of the division of epidemiology at UCSD, Elizabeth Barrett-Connor. As she explains it, the Garlands were simply butting up against the realities of medical research. "They weren't taken seriously," she says, "but why would you take them seriously? Hypotheses are generated all the time. Why would you take that one seriously, as opposed to the other fifty that year? Besides, the whole arena of vitamin deficiency has been and still is incredibly controversial."

Still, she was sympathetic to the brothers' efforts, while concerned that they had become too personally involved in proving their case. "If you have an excellent and original idea, which this was," she says, "it's potentially dangerous to become so enamored of it that you forget that the task of a scientist is to prove your hypothesis is wrong, not to prove it's right. Real science proceeds when people set out to challenge their hypotheses."

So she approached a Chicago epidemiologist named Richard Shekelle. Shekelle was sitting on a gold mine, a huge compilation of information taken from 2,000 men who worked at the Western Electric Company in Chicago in the 1950s. These men had answered an exhaustive question-naire about their eating habits and had then been examined periodically for 20 years afterward. The idea had been to explore the relationship between diet and heart disease, but the men were also scrutinized for other problems, including colon cancer.

"These people had been asked, 'How much beef do you eat every day? How many potatoes do you eat every day?'" says Frank. "The interviewers would write it all down and store it away. It was powerful information, because it was done *before* the people became ill. Much more powerful than if we had people here with colon cancer and said, 'What did you eat before you got sick?' They'd tell us what they wanted us to hear."

Over drinks at a 1983 Lake Tahoe conference, Barrett-Connor and another UCSD colleague, Michael Criqui, persuaded Shekelle to let the Garlands comb his data. Here was an opportunity to test their theory against the fates of individual people. It revitalized the Garland brothers.

"We were stuck," recalls Cedric. "We were at the point where we had exhausted all our resources. It was a great relief to know that they would be willing to work with us."

The work lasted for two years. "We might have looked just at vitamin D, since that was the hypothesis," says Cedric, "but we decided to look at everything, just in case. We looked at all the vitamins, all the minerals. We looked at protein, fat, calories, fiber. Nothing had an effect on colon cancer."

Nothing, that is, except vitamin D—and as it turned out, calcium. The Garlands determined that the risk of contracting colon cancer for men who consumed more than 150 international units (IU) of vitamin D per day (the recommended daily allowance for adults is 200 IU) was half that of men taking in less than 150 IU. And when it came to calcium, the results were even more striking. As the consumption of calcium increased, the risk of colon cancer went down, until the men taking in the most calcium had less than a quarter of the risk of those consuming the least. (The results were no surprise, as vitamin D and calcium are closely linked. One of the major functions of vitamin D is to facilitate the body's uptake of calcium.)

The Garlands felt vindicated. Says Cedric, "It turned out to be one of those studies in which there is no way that you could make the findings go away—no matter how much some members of the group might have favored that happening. They knew that because we believed in this hypothesis so strongly, it would have been easy for us to use tunnel vision and just look for what we wanted to see. But nothing except vitamin D and calcium showed an effect on colon cancer."

Anticipating that the study would generate at least a semblance of their own enthusiasm, the scientists published the findings in the influential British medical journal the *Lancet* in 1985. The response wasn't what they had imagined. "We got some reprint requests," recalls Cedric. "We traveled around quite a bit and presented the paper at meetings. Nobody," he says wanly, "had any strong objections."

Neither did anyone offer strong support. In fact, the most interested reaction involved vitamin D only secondarily: the Garlands were invited to present their findings at the Memorial Sloan–Kettering Cancer Center in New York, where an investigation into the relationship between calcium and cancer was already under way.

If the vitamin D findings didn't provoke much excitement, however, perhaps the limitations of the study were at fault. As the questionnaires only chronicled what people ate, they couldn't take into account the original focus of the Garlands' investigation—sunlight. "Most vitamin D comes from exposure to sun," says Frank. "Dietary vitamin D isn't going to give you the whole picture."

The Western Electric study, then, was only a start. If the brothers could now figure out a way to nail down sunlight-derived vitamin D as they had nailed down dietary vitamin D, they'd really have something. The question was, how to do it?

Perhaps Operation Clue was the ticket. In 1974, their first year at Johns Hopkins, the Garlands had spent the summer in the town of Hagerstown, Maryland, helping to set up an ambitious Hopkins blood-banking project called Operation Clue. "Our group rented five trailers and hired five nurses and ten assistants," recalls Cedric. "They went out to the Mack Truck factory, shopping centers, schools, collecting blood from 25,620 donors. And it was all frozen."

"The purpose," says Frank, "was to put the blood away for questions that had not yet been posed."

The brothers hadn't forgotten about all that blood, but they hadn't felt justified in using it. "It was a valuable and limited resource," says Cedric. "You had to present a pretty strong case before you'd be allowed to unfreeze the samples. We hadn't quite achieved that yet." Not until the Western Electric study, that is. Now the brothers were ready to make their case.

The crux was that you could measure the amount of vitamin D in blood—even in blood that had been frozen over a decade. Because the donors had been tracked for 8 years afterward, the brothers could compare the amount of vitamin D in the blood of those who later suffered colon cancer with the amount in the blood of those who did not. And in contrast to the Western Electric study, it would be possible to measure both dietary vitamin D *and* vitamin D generated from the sun—the amount in the blood reflected both sources.

It was a stronger study for other reasons as well. "It's one of the most absolute, most powerful studies in epidemiology," says Frank, "because it's not subject to the problems of a dietary study—no biases on the part of the interviewer, no faulty memories from the person being interviewed. This is blood from healthy people. You take something that can be measured in a laboratory, and you see who gets sick afterward." For the Garlands, perhaps, Operation Clue just might live up to its name.

The results were spectacular. People whose blood contained lots of vitamin D contracted colon cancer only one-fifth as often as people with low levels in their blood. In other words, high levels of vitamin D signified an 80 percent reduction in the incidence of colon cancer. And this was vitamin D from diet *and* sunlight.

The information appeared in the *Lancet* in November 1989. This time it did not go unnoticed. After a few months Cedric's phone rang. "It was from the National Cancer Institute," he recalls. " 'A director's seminar has been set up for you and your colleagues to present your results. Get your tickets, fly out, and give your presentation.' "

"We were thrilled," says Frank. "Director's seminars are special events. They're action oriented. Something is supposed to come out of the meeting that will carry on." And if the NCI and its parent organization, the National Institutes of Health, wanted to carry on, they certainly had the resources to do so.

So on April 26, 1991, Cedric, Frank, and Edward Gorham, Cedric's graduate student and the brothers' recent collaborator, hopped a plane for Washington, D.C. And to a group of thirty people, including the director of the NCI and the heads of its divisions, they told their story, beginning with a map like the one Cedric and Frank had seen flashed on the screen at Johns Hopkins years before.

It was the climax of their efforts—high-level recognition that the vitamin D–colon cancer connection was indeed real and potentially important. As of this writing, many months later, the trio still recall the session with wonder, as you might a powerful dream.

Ed: "We were expecting trial by fire. We were ready for it, but it didn't happen."

Cedric: "We anticipated all sorts of questions about our hypothesis or evidence. But all they wanted to know was our assessment of an optimal dose [the Garlands recommend a daily intake of 400 IU of vitamin D and 1200 milligrams of calcium] and suggestions of what to do. We recommended a clinical trial."

Frank: "That's when we became almost like outsiders. They got into this deep discussion of how they were going to go about involving these ideas in their programs."

Cedric: "The whole thing lasted a little over two hours."

Ed (to a burst of laughter from the brothers): "Afterward, walking out of the NCI, we felt that if we got hit by a truck at least now the work would keep going."

POSTSCRIPT

Last winter the National Institutes of Health announced a 10-year, $500-million study involving a clinical trial of some 70,000 women to test, among other things, if vitamin D and calcium prevent colon cancer. It will be the trial that the Garlands have been aiming toward for so long, the intervention experiment that is the climax of any good epidemiological detective story.

No one claims that the brothers' work instigated the study—"It was already in the pipeline," says Peter Greenwald, the director of the Division of Cancer Prevention and Control at the NCI, "but mainly to see if calcium might prevent osteoporosis"—but it enlarged the focus of those involved, convincing them to incorporate vitamin D and colon cancer into the trial. "The Garlands reviewed their evidence at the director's seminar," says Greenwald. "They've got a good handle on it. We think it's important research—that's why we're encouraging this study to find a clear answer."

"We're delighted beyond words," says Cedric. "Finally the scientific community has enough confidence in the work to give it an opportunity to prevent colon cancer." Meanwhile, the Garlands and Gorham seem to be finding a similar connection between sunlight, vitamin D, and breast cancer and are looking into ovarian cancer as well. Perhaps now, fourteen years after their initial insight, they can begin to pull the handle off the pump.

Peering into Shadows

Thomas Levenson

In the last chapter of his third book of essays, Montaigne wrote, "There is no desire more natural than the desire for knowledge. We try every means that may lead us to it. When reason fails us, we make use of experience . . ."

Experience is a "feebler and less worthy means," Montaigne goes on to say, but ignore any road to truth at your own peril.

Experience, an experience, is something we have, that happens to us, that we can possess. Truth becomes property, the object of our desire. But there are events that one cannot grasp, seize, own. In the summer of 1991 a total eclipse of the Sun occurred along a track that stretched from the central Pacific to Brazil.

Most of the twenty thousand people who came to the Big Island of Hawaii to view the eclipse were drawn by the promise of near-perfect viewing on the lava fields on the northwestern coast of the island. There, with luck, it would be possible to peer out over the broad ocean and witness, just before the moment of totality, a wall of shadow racing across the water. Annie Dillard has described the onslaught of the Moon's shadow during an eclipse seen in Washington state in 1979: "We no sooner saw it than it was upon us, like thunder. It roared up the valley. It slammed our hill and knocked us out . . . If you think very fast you may have time to think, 'Soon it will hit my brain.' "

For Dillard, the sight of that shadow was literally a revelation: direct confirmation of what to her had previously been only words, ideas. She wrote, "This was the universe about which we have read so much and never before felt: the universe as a clockwork of loose spheres . . ."

When it came my turn to travel to an eclipse, that image of the clockwork universe was much in my mind. The eclipse of 1991 was unique in the history of scientific eclipse observation, made so by a coincidence of geography: the presence in the eclipse's path of a mountain, Mauna Kea, at 13,800 feet the highest point in the Pacific.

Seven large and two small structures occupy Mauna Kea's summit, the grandest of them graceful white, vaulting domes, all of them housing telescopes taking advantage of one of the world's best sites for the study of deep space. The 1991 eclipse was the first to occur over a major modern observatory, and it offered solar scientists their first chance ever to use sophisticated telescopes to explore the Sun. This was the eclipse for those who created the image of the clockwork universe.

The solar scientists' experiments were experience systematized, experience sought and structured and fraught with the possibility of discovery. Eclipses illuminate the world beneath them in a glow that is unlike any other, silver, thin, and brittle. Science maps the world of experience; from its origins, it has constructed stories that can anchor us within that world. The shuttered light of the eclipse of July 11, 1991, gave a sudden glimpse of how this can be done and why we feel compelled to do it.

1. DAWN

The air is thin on top of Mauna Kea. At the summit, atmospheric pressure runs just under 60 percent of its value at sea level. The lack of oxygen makes walking on level ground tough, to say nothing of the effort required to climb a flight of stairs. Old Mauna Kea hands have the rhythm down: step, breathe; step, breathe; step, breathe. There are compensations, raptures of the heights. Hypoxia produces a kind of euphoria, giddiness. But it is hard to sleep at such an altitude. Unconscious fears of breathlessness, perhaps, cause one to toss and turn, to wake and sleep and wake again. Dreams come constantly, vivid, distracting, elusive, ultimately exhausting. No one (officially) sleeps at the summit.

Eclipse day began at the dormitories constructed at the 9,000-foot level on the flanks of the mountain, just about where the scrub foliage gives up and pure lava cinders take over. The facility had run out of beds two nights before, but on clear nights it was a pleasure to sleep outside: the old Moon waned fast just before the eclipse, and the stars crowded together in bursts of shapes and lights impossible to see at sea level, through the murk of another mile and a half of air. But the weather on the last night was dicey, with patches of cloud and a hint of dampness in the air. I bedded down in the backseat of a Chevrolet and woke at 2:30, when a trickle of rainwater dribbled down one window and splashed onto my face.

Rush hour had already begun. The timing of the eclipse—beginning at 6:28 A.M., with totality coming an hour later—dictated a miserable schedule in which sleep, for a week, was compressed into ever narrower bounds, the few hours between late evening and the middle of the night. On eclipse morning, the run up the mountain began before midnight, as teams of astronomers raced to cram in a last rehearsal. The clouds bottomed at about 9,500 feet and broke abruptly somewhere above the 11,000-foot mark. The sky was dark, clear, and the North Star—essential for the alignment of certain telescope mounts—was clearly visible.

The use of the North Star reflects one of the pleasures of astronomy. Unlike some sciences, in which new discoveries render irrelevant (or at least seem to do so) the apparatus of previous constructions, astronomy retains much of the scaffolding upon which it was founded. Polaris guided travelers innocent of any modern definition of a star; it still provides the observer with a useful point of reference. But astronomers themselves have some tendency to forget where they came from. Eclipses are now actually out of fashion among serious researchers. The 1991 eclipse was an abberation because of the opportunity created by the presence of the Mauna Kea observatory in its path. Mostly, though, astronomers see eclipses as a closed book. The orbital geometry that gives rise to the perfect overlap of Sun and Moon is a happy coincidence; the motions are governed by Newton's laws; the Sun itself, near at hand and (for our sakes, fortunately) a rather dull and quiescent star, is far less compelling than the bestiary of the night sky: black holes, quasars, and the rest.

It wasn't always so. Eclipses have a glorious history. The effort to predict them is as old as observation of the sky. In legend, more than four thousand years ago the Chinese emperor executed two astronomers who had befuddled themselves with wine the night before an eclipse and missed the crucial event altogether. Babylonian astronomers discovered fundamental rhythms in the cycle of eclipses using records kept from 3000 to 2000 B.C. Greek observers were capable of fairly precise predictions. Thales of Miletus conceived the notion that all the complexity of nature should be explained by the minimum number of hypotheses; his ideas underlie all theories of matter that have come after him. This ancestor of modern science predicted the eclipse of 585 B.C., whose onset so startled the less well-informed armies of the Medes and Lydians that the two nations concluded a peace treaty on the battlefield.

The Newtonian revolution put the problem of predicting eclipses to rest, in theory. Newton's laws provide the mathematical apparatus for calculating the courses of Sun, Earth, and Moon, and Edmond Halley discovered the slight acceleration in the Moon's pace as it orbits around the Earth that served as the first correction to be applied to any given eclipse calculation; Halley was able to put his work to the test in 1715, when he stationed observers along the predicted track of an eclipse across the south of England to report any discrepancies of time or path. The experiment was repeated in New York City in 1925, when the southern edge of the path of totality cut across Manhattan. Consolidated Edison employees were stationed on rooftops on every block heading north on Broadway to identify the precise boundary of the eclipse—between Ninety-sixth and Ninety-seventh streets, as it turned out.

In fact, to this day, absolute precision is impossible. The Earth is not a perfect sphere, and its irregularity affects its course around the Sun, while the interaction of all the bodies in the Solar System produces a constant jiggle in the orbits of the all planets. The changes are slight—

any year is pretty much the same length as any other—but even a small wobble matters to aircraft and ships at sea, for the changes cause a clock set by the rise and fall of the stars to run slightly awry, creating navigation errors that can extend to miles. (Standard time is reset occasionally by an international committee checking the stars against an atomic clock: we still guide ourselves by the light of a beacon star.) The bob and weave of Earth and Moon also place an ultimate limit to the accuracy with which we can predict where both will be with respect to the Sun even a short distance in the future. Calculations of the onset of eclipse at Mauna Kea turned out to miss by as much as ten seconds.

Such small errors don't matter much. No one runs the risk of being beheaded for fluctuations easily spotted and swiftly accounted for. But the scientists came to Mauna Kea on a mission not too different from that of Con Ed's finest back in 1925. Hal Zirin came from Caltech to Mauna Kea simply to record the radio emissions of a tiny region of the solar atmosphere just above the surface of the Sun. That radio signal contains clues about the temperature and density of the inner solar atmosphere, but it is only measurable during an eclipse. Ordinarily the bright glare of full daylight obscures any emissions from the neighborhood immediately above the surface of the Sun—so the Mauna Kea eclipse represented an unique opportunity for Zirin. Yet Zirin was after facts here, nothing more. "All we'll know, then," he said, "is what exists on the Sun. But then we can go back and try to explain it. Too many times we are explaining things that we don't even know exist."

That's an old-fashioned view of what scientists do, a venerable one. Newton himself gave it his seal of approval, writing at the conclusion of *Opticks,* "The investigation of difficult things by the method of analysis . . . consists in making experiments and observations, and in drawing general conclusions from them by induction, and admitting of no objections against the conclusions, but such as are taken from experiments or other certain truths." Science advances as observations multiply—as machines like Zirin's radio telescope extend our senses, allow us to "see" ever more detail in nature. Zirin's experiment was an example of what those who created the scientific revolution imagined science to be: an accumulation of unequivocal facts of nature from which can be drawn the patterned laws of science.

These are the facts as they began to accumulate on the day of the eclipse: by a little after 5:00 in the morning, eclipse day, the sky had begun to lighten off to the east. I stood outside with some of the astronomers, next to one of the large domes, staring down at a field of cinder cones perched on a plateau several hundred feet below the summit. Fog twisted in and around the cones, sometimes curling out of the old, dead craters, the ghosts of smoke that rose thousands of years ago. On the horizon colors brightened and spread, north and south, purple coalescing into orange, then yellow. I watched the swirls of cloud, the colors, the breath of wind rising as the air warmed over the lip of the summit ridge. The minutes passed, leaving behind them the moment when the Sun was supposed to appear. We waited. One of the astronomers wandered along the ridge and raised his arms to command the rising of the Sun. The clouds below extended to the edge of sight, delaying the break of day. Suddenly, the shadow of the mountain shot west for miles, with such clarity that one could discern the outlines of the domes of the telescopes against the bright white screen of the overcast below. It was the morning of the eclipse, dawn, 5:52 A.M.

2. FIRST CONTACT

The tools astronomers build use something of the old sympathetic magic, likeness revealing the nature of likeness. Mechanical orreries, clockwork models of the solar system, traced the evolving patterns in the dance of the planets around the Sun, and in their gears and linkages they supplied the metaphor with which their age painted the universe. The great observatories also weave a picture of the sky into their fabric, as you can see if you look past the clutter. With first light on eclipse day, the crowd waiting for the Sun broke up on the mountain's edge, and the people trailed into the buildings where their machines waited. I followed, and I took one last look at the telescope within the dome before burying myself in the control room.

That particular instrument was a 3.6-meter optical telescope, the seventh largest in the world. Its scaffolding and mirror were held within a giant yellow U-shaped yolk. At rest, it pointed straight up into the heavens. From time to time a voice would come over the loudspeaker and machinery would move, motion that mapped the sky. An observatory's dome is simply the dome of the sky with the rain kept out. In an ordinary night of observing, when an astronomer wishes to remain focused on a single object, the dome turns slowly, revolving at a rate of one revolution per twenty-four hours, countering, step for step, the rotation of the Earth. Every use of a giant telescope recalls Galileo's stand: the heavens remain in place; we move.

Individual experiments contain the same archaeology, the same lines of connection between what we thought and what we think. One team at Mauna Kea, using a 2.2-meter optical telescope, was planning to take pictures of the solar corona, the Sun's outer atmosphere, which would be visible only during the total phase of the eclipse. In itself that was as old an aim as the viewing of eclipse, but the group added one twist: they planned to take pictures in the pure light of a single color, a green light with a wavelength of 5,303 angstroms, light with a pedigree and a deep family connection to eclipses.

The historian Owen Gingerich of Harvard related much of the tale. As participants remember it, the story began with a fire. One evening, the German scientists Kirchoff and Bunsen saw a fire burning in some distant buildings. They knew already that chemical elements, when burned, give off characteristic colors of light, called their *spectra*. (Fireworks makers use the color differences to achieve their effects—copper lends a green color to a flame, for example, and sodium burns a bright yellow, the yellow of sodium streetlights. That yellow is only the brightest color that sodium emits. Sodium, like the other elements, can produce a range of colors of light, and it was those spectra that Kirchoff and Bunsen studied.) But when the two men saw that city fire, the story goes, they turned their instruments on that source of light. They detected spectra in the flames—and thus discovered a method for investigating the chemical composition of firey places at a distance.

Next they looked upward, and identified over a dozen different chemical elements burning in the flames of the Sun. This discovery, in 1851, was one of the landmarks in the history of science: it was the first definitive evidence that objects beyond this planet are built from the same stuff as everything on Earth. It also drove astronomers to apply instruments called *spectroscopes* to the viewing of eclipses. In 1868, Pierre Jules César Janssen of France looked at the bright flames of solar prominences at an eclipse in India and saw spectral patterns never before observed. It was the trace of a previously unknown element, which was to be named *helium,* after Helios, the sun.

In the next year, the leading American eclipse scientist, Charles Young, took his spectroscope to an eclipse in Iowa and detected in the corona a bright green signal, also new, unfamiliar. The element with which it was associated was dubbed *coronium,* but no periodic table today contains mention of the substance.

Eclipse observations seem to turn up ghosts. The event itself is so spectacular that even seasoned observers forget themselves, lose hold of their senses for a moment. Wobbles in the orbit of Mercury persuaded some 19th century astronomers that another planet lurked still closer to the Sun, as similar irregularities in the orbit of Uranus had led to the discovery of Neptune. At the eclipse of 1878, one astronomer persuaded himself that he had seen a speck in the sky and announced the discovery of a new planet, Vulcan. And Young himself was so stunned by the beauty of the 1869 eclipse that he forgot to look through his spectroscope at the moment crucial to his main experiment; he cursed himself afterward, noting the existence of the green line as an afterthought.

Vulcan evaporated when Albert Einstein accounted for Mercury's orbit with his general theory of relativity, but the green line is real. With the development of the new physics in the twenties and thirties, the meaning of that green line became clear: it was the trace not of a new element but of an old one in an unfamiliar form—it was the signature of iron atoms that had had thirteen electrons stripped away. That happens only at temperatures of about 2,000,000° Kelvin. The green line thus became a thermometer, a probe that revealed that the solar atmosphere simmers at 2,000,000 degrees, while temperatures at the surface of the Sun are about 6,000 degrees. What drives portions of the corona to such extreme temperatures remains a mystery. The team at the 2.2-meter telescope at Mauna Kea planned to use that bright green light to search for the physical processes—explosions, magnetic storms—that could burn brightly enough to smelt iron into flames of vivid green. It was to be the same experiment as the one Charles Young performed—or rather, it was the descendent, the lineal heir: Young looked at the eclipse to learn of what the Sun was made; the scientists in 1991 knew what was there and sought an explanation of how it got that way.

Young stood out in a field to watch the eclipse of 1869. A photograph from his expedition shows a landscape as flat as a dance floor. There is a large tree on the left, and a kind of makeshift three-walled tent, and Young himself sitting at his spectroscope off to the right. As the morning of the eclipse of 1991 moved on past 6:00 A.M. I found myself in a windowless control room four floors up. All such places look alike, with their consoles, banks of dials, computer screens. We had a video hookup to the outside, showing the Sun like a white cutout circle on a small monochrome monitor.

In Galileo's time (and, still, in our own) telescopes and scientific instruments in general were thought of as machines that simply extended human senses, bringing the distant near and revealing mountains on the Moon, satellites around Jupiter, spots on the Sun. But in fact now, as in Galileo's day, the extension of sight goes beyond seeing more of the same; it alters the nature of what is seen. The telescopes used during the 1991 eclipse expanded upon human senses, taking us into realms beyond our experience. Facts are not simply facts—the green line is not merely green, nor is it the spoor of coronium. The meaning of a green trace on a photographic plate changes as our ideas change.

And in this expansion of sense (and experience, in fact, for we do see that green light and know its meaning) we reap some consequences, for we must dissect the image of the Sun to reach its parts. Staring at a television screen, one is unlikely to imitate Young, to abandon reason and stare upward at the crucial moment. At 6:28 we waited. The telescope operator counted down, then shouted, "Mark!" We peered at the white circle on the monitor. Eight, ten seconds later, there it was: a tiny curved bite out of the top of the ball. Someone announced over the loudspeaker, "First contact confirmed." Down on the coast—40 miles, or 13,000 feet, away—the crowds cheered, and a few screamed. In the control room, one of the scientists said, "I guess we got the right day." We had been holding our breath; now we exhaled.

A minute or two later, I ducked out and found my way to the catwalk. The sky was bright blue. The clouds stretched out below, moving and shuddering with the morning updrafts. The light seemed perfectly normal; it was a beautiful day. And then I looked up through a filter and saw the Sun with just that one nibble missing. It was real. It had begun.

3. SECOND CONTACT

It took the Moon about an hour from first contact to completely eclipse the Sun. That hour was elastic, stretching and shrinking at every passing instant. In Hawaii, the Moon did not seem to move at all. Then I would look away, or go for coffee, or just walk around the circumference of the dome for luck—and the Moon would chop away at the Sun. Time did not flow; it shuddered. An eclipse upsets all the normal points of reference.

I went back out on the catwalk just about 7:00, with half an hour still to go before second contact, the onset of totality. It was a moment teetering between the appearance of normality and the territory of the completely strange. With about half the Sun gone, the light of day still shone. The people working on the ground beneath me continued to move through three dimensions, space and distance maintaining their ordinary relations. But as I waited, leaning against the rail, the colors shifted just a little, and the world slipped.

The sky began to darken—first simply turning a deeper blue. At the same time, the light seemed to thin. It became stiletto tipped, cutting ever-finer edges. The man standing next to me began to pale, just a little, and his shadow started to fade. Above us, the Sun was about two-thirds

—

covered. I walked around the catwalk to the hatchway where the portable experiments were set up, and I watched perhaps a dozen scientists attending to their machines. Every few seconds one of them would grab another and they would stop and look. People are unbelievably still at such moments, frozen, fixed within the play of alien light. Some seemed almost to shake off bonds as they returned to work, checking the alignment of their portable telescopes and cameras.

One of the Mauna Kea set-ups was dedicated to the simplest of observations, recording images of the star field that would appear around the Sun during total eclipse. It was a sentimental experiment, homage to the most famous eclipse expedition of them all, when in 1919 Sir Arthur Eddington led his team to Principe Island, off West Africa. The expedition nearly ended in failure. The weather was miserable, with the eclipse ducking in and out of the clouds. At the start of totality, no stars were visible to the naked eye. The astronomers soldiered on anyway and made sixteen photographs. The clouds seemed to lift toward the end of the total phase, and Eddington obtained one picture with what he termed "fairly good images of five stars." It was a photograph that transformed the cosmos.

Or so said the *London Times* on November 7, 1919, with headlines that read "Revolution in Science: New Theory of the Universe, Newtonian Ideas Overthrown." Eddington compared the positions of the stars he saw during the eclipse to images of the same stars taken at night, and he found that there had been a change. By all appearance, the stars in their courses had moved.

Or rather, light had moved, bent around the gravitational pull of the Sun, producing an apparent displacement of the stars closest to the Sun in the eclipse-darkened sky relative to those farther away. Newton himself had suggested that the Sun might bend light, and his mathematical account of gravity generated a number for the amount of deflection. But that value for the bending of light was never confirmed.

Albert Einstein's approach to the problem of gravity began in 1907, when, as he later said, he saw a workman fall from a roof. He wrestled with the problem on and off for several years, in 1911 coming up with a mathematical theory of gravity that produced a number for the deflection of starlight that was the same as Newton's value. At that point, Einstein got lucky: the first expedition to test his idea at an eclipse was rained out in Argentina in 1912. A German eclipse expedition did set out for the Crimea in 1914, only to be taken prisoner at the start of World War I.

Which was good for Einstein, if not for the expedition party, because his original idea was wrong. Einstein's ultimate theory of gravity, the general theory of relativity, was finally completed in Berlin during the late autumn of 1915. It produced a value twice as large as the Newtonian number for the bending of light around the "curve" of space created by the gravitational field of a massive object like the Sun.

Einstein's theory passed its first test when it explained the variations in Mercury's orbit that had suggested the existence of the planet Vulcan to previous generations of eclipse observers. The next major test was Eddington's attempt to measure the true value for the deflection of starlight at an eclipse. Einstein himself professed a lack of concern whether the Eddington photographs would turn up the "wrong" number, saying, "Then I would have been sorry for the dear Lord. The theory is correct." But when he heard unofficially that Eddington had confirmed his ideas, he gave a hint of his true feeling in a postcard home: "Dearest Mother, joyous news today . . ." The Royal Society held a meeting on November 6, 1919, to discuss general relativity, and they concluded, "The result is one of the highest achievements of human thought . . . It is not the discovery of an outlying island but of a whole continent of scientific ideas."

With twenty minutes to go to totality at my eclipse, I returned to the observatory control room. The television picture was getting fuzzy. The abnormally high clouds in the valley beneath us had been heating up since dawn, and they rose as they warmed. Some fog was beginning to drift upward across the face of the Sun. We had this connection too with Eddington—like him, we raced the clouds for sight of the Sun.

Since Eddington left Principe, general relativity and its description of the large-scale structure of the universe have been confirmed with ever-increasing precision. After the Second World War, eclipse measurements gave way to the study of the way radar bounces off Venus, and to other, still more esoteric tests, all of which Einstein's theory of gravity has withstood. Any scientists repeating Eddington's original experiment in 1991 could not have done so in any hope of breaking new ground.

I don't know exactly why they did attempt their re-creation. In part, it was simply a test of virtuouso skill. The measurements are technically difficult, and reexamination of Eddington's plates has turned up errors larger than were noticed at the time. And perhaps someone simply wanted an excuse to come to an eclipse. But there was more to it than that, at least for me.

Alfred North Whitehead took part in the Royal Society meeting that reviewed the Principe results, and his memory hints at a reason. He wrote, "The whole atmosphere of tense interest was exactly that of the Greek drama. We were the chorus commentating on the decree of destiny as disclosed in the development of a supreme incident . . . The essence of dramatic tragedy is not unhappiness. It resided in the remorseless working of things . . . This remorseless inevitableness is what pervades scientific thought. The laws of physics are the decrees of fate."

Science, for the scientist, is often mundane. The astronomers at Mauna Kea soldered wires, fixed balky instruments, loaded film. Some of them had traveled thousands of miles to remain stuck within a windowless room, staring at computer screens throughout the eclipse. The observations themselves were not of the Eddington variety—elegant, economical tests of whole new world systems—but rather concerned themselves with much smaller questions, the ordinary questions of everyday science.

But Whitehead recognized the motive that lies within the urge to know. I waited for the eclipse, for the echo of the encounter Eddington had, seeking that sudden view of the "remorseless working of things." Science is not simply what scientists do. It is the pattern we impose upon our world; it is the story we tell ourselves, a story that comforts and terrifies, that connects the supreme incidents of experience to the decrees of fate. That is why we still travel to eclipses: to see truly what exists, what can occur, what continents we inhabit.

—

Fifteen minutes before totality, I stepped outside once more. The clouds had risen farther and piled themselves against the summit cone. Wisps blew up and hid the Sun from time to time. The light had turned sharply. Colors on the ground lost luster, became flat, while the sky, at the moments the clouds broke, took on an unnatural depth, seemed to recede in a blue deepest near the shrouded Sun. The clouds surged upward one last time, blocking the eclipse entirely. But then, as the Moon continued to cover the Sun, the temperature began to drop—no more than a couple of degrees or three—but enough to settle the fog. Perhaps a dozen minutes before totality, the clouds fell below the summit and the Sun shone past the edge of the Moon. The sky transformed itself. Sunset colors of orange and purple spread in the middle of the sky. To the west a kind of gloom gathered, indistinct, spreading north and south. I held up my hand, and the last light of the Sun seemed to wash through it—my skin had the look of old glass. Back inside the control room, one man spoke: "Wait, wait, wait," and then, "Okay, you can go," and then someone told us over the loudspeaker, "We have total eclipse on Mauna Kea."

I can describe what I saw when I ran outside again. Instead of the Sun, there was a black circle that looked like a hole punched in the sky, surrounded by a white crown—the corona, which was wispy, extending in jagged petals to the left and right of the Sun. Top and bottom, huge orange loops of flame curled above the edge of the Moon. I could see from side to side of the eclipse—Mauna Kea's summit was high enough that I could look toward the horizon and clearly see both the inverted cone of darkness that was the Moon's shadow and the bright light of day beyond. I saw— it doesn't matter, for what I saw I saw, and I can describe that sight, but I cannot convey the vision. I saw the impossible, the Sun stolen from the sky, the shadow of the Moon, day become night. I saw a curtain ripped aside and behind it, a universe I had not known.

4. Afterward

Time wobbled again during totality. We just looked, still. It was impossible to tell how long we stared; we were outside the flow of time and, peering upward, seemed locked in a single moment. In an instant, the total eclipse ended. The top of the Sun peeked out from behind the Moon. The corona disappeared. We could see the moon's shadow, a great bowl shape, heading east, back out to sea. And it was over. The partial phase would continue for another hour, but no one looked up anymore. The clouds, released from the chill of darkness, rose again, and rain fell later that morning. There was a press conference at which the scientists reported on their experiments, and then we ate a catered lunch and drank from a little fountain flowing with white wine. The airport at Kona was packed with departures that night, and throughout the next day, too.

Shortly after I returned to the mainland I reread a favorite poem, one by by Wallace Stevens called "Variations on a Summer Day." In it I was caught by first one line: "An exercise in viewing the world," and then another: "To change nature, not merely to change ideas."

As of this writing, the results of the eclipse experiments are being analyzed. The team trying to photograph the bending of starlight failed, as high cirrus clouds blocked their view of the most important stars, but the large telescopes were able to see past the high haze into the solar atmosphere itself. One team photographed the smallest structures ever observed in the corona, tiny magnetic explosions; another tracked an anomalous spectral line to the precise location above the Sun where the magnesium reactions that create it take place. The green-line team made a movie of the hot corona. And measurements of radio emissions near the edge of the Sun suggest that the inner solar atmosphere is four times denser than previously believed.

"An exercise in viewing the world. . . ." The eclipse experiments were such an exercise, as is all of experimental science—and so is, or could be, any waking moment. And then, "To change nature, not merely to change ideas." The eclipse experiments of 1991 recapitulated three hundred years' worth of exploring the territory revealed at the overlap of Sun and Moon. We have sought to extend the reach of our senses, to expand them, to discover, sometimes, the Royal Academy's "whole new continents." Our picture of the world—nature as we understand it—has been transformed as our ideas have changed over those three hundred years. But most of all, in the act of knowing, through the experiences which lead to knowledge, we ourselves are being transformed.

Since the morning of the eclipse, I have come to examine the sky constantly. I watch the Moon shift its phases; I seek the North Star every clear night; I take note of the shifting track of the Sun as we swing through the seasons. To witness an eclipse is to place oneself in the path of nature; the nature that was changed on July 11, 1991, was my own.

REDUCTIO EXPANSIOQUE AD ABSURDUM

DOUGLAS R. HOFSTADTER

In memory of Kees Boeke, author of Cosmic View

I. REDUCTIO

Scale One

She swung around the bend—a little too fast, she knew, but she needed to get there quickly. Her headlights bounced across the snow, revealing corrugated mounds several feet high at the road's edge, lining the fields. She stepped on the accelerator again and gunned the car down a straight stretch. Ahead she saw a haystack and realized another bend was coming. Timing it carefully, she put on the brakes and careened around the bend to the left, almost but not quite skidding out of control. Risky? She knew it was, yet it was exhilarating at the same time—living at the edge of the abyss, tempting fate.

Her headlights' glare caught something—something moving—something darting ahead. What's that? She slammed on the brakes. There was a loud noise as a deer smashed into her car's front end, bumper then hood then windshield, all in a fraction of a second, all blindingly fast, leaving her windshield a shattered web of white with a large hole in it where the antlers and then the head had pierced. The car was still going forty miles an hour when the deer's antlers entered her skull.

Scale Two

A piece of antler horn was moving slowly through the air, first left, then right, then bobbing up and down a bit. A suddenly loud-growing roar was accompanied by a freezing-in-place of the bit of horn. Only half a second later, there was an enormously loud sound and the piece of horn jerked wildly. Suddenly it was being dragged along at forty miles an hour and in a strange path.

The piece of horn bounced down and slid across a cold, painted metal surface and in a tenth of a second encountered a thick sheet of glass. By itself, it would just have glanced off the glass, but it was part of a much larger piece of horn that was attached to a large massive object that was also being propelled in a strange, violent path toward the glass. So the horn, when it hit the glass, exerted much force on the sheet, and in a thousandth of a second, the sheet yielded and then cracked, with glassy splinters moving away in several directions at once from the hole the horn was opening up, while a little farther away, rays of white were shooting through the glass and forming an intricate, dense, crisscrossing web of fracture lines. The end of the piece of horn was now quite blunted, but it continued to move in approximately the same trajectory, as it was still attached to the larger body behind it, close on its heels, moving toward the glass, where it would soon open up a much larger gaping hole.

The horn fragment was now in open air, not too many inches from a suspended mirror. It was moving at about thirty-five miles an hour relative to the mirror, having been slightly slowed down by the collision with the glass. Now it quickly crossed a one-foot stretch of empty space, and then encountered a piece of softer material—a light-colored, warm membrane, which it instantaneously pierced. This membrane covered a heavier piece of bone a quarter of an inch thick, which offered as much resistance as the glass had. It was moving toward the mirror at about five miles an hour, so the horn entered the bone at nearly forty miles an hour. Again the large mass behind it impelled it forward and the bone gave way, splintering and cracking audibly.

Beyond the bone there was a very soft mass of sticky substance much warmer than the outside air. The piece of horn moved swiftly into that substance, which was so viscous that it soon impeded the horn's further motion after a couple of inches. The horn came to rest lodged in a hunk of red, oozing, spongy matter, warm and pulsing.

Scale Three

A packet of neurons was firing away rhythmically, receiving and sending pulses every few milliseconds. Its many thousands of neurons were all engaged in a collective mode, like birds in a flock, so that when one altered its activity, all the others were quickly affected and a ripple would pass through the many neurons, putting the team into a slightly different collective mode. Every eight milliseconds, roughly, the periodic pulsing of this particular mode took place. As regular as an oscilloscope trace of a slowly changing sound, the pattern repeated. Each cycle was just barely distinguishable from the preceding one—an adiabatically changing pattern.

Then an anomalous pulse train came in from another team of neurons that didn't usually communicate with this team. Its pulses were a little faster than normal, and they had an imperative quality to them. The pulse pattern was thus disrupted from its slowly drifting, adiabatically shifting periodic firing pattern, and it made a rather abrupt transition into a different mode.

In this mode it carried out about a dozen complete cycles when another disruptive event took place: several hundred of its neurons were severed from the main body (which contained around eight thousand neurons altogether), and no more influence from them was felt. Thus the firing pattern was again altered and became less periodic, for the usual stabilizing effect of the closed system was gone. The pulsing was a little irregular but continued for many thousands of periods longer. Then it slowly ebbed, and there was a gradual cessation of firing.

Scale Four

A retinal cell was firing away quite intensely. Its job was to respond to a brightness gradient oriented at about sixty degrees from the horizontal, relative to the straight-ahead gaze of the eye to which it belonged (although it knew nothing of eyes or their gazes). The cell was firing fairly rapidly because there was, in fact, just such a brightness gradient, a fairly strong one, at forty degrees, which was close enough to sixty degrees to induce a pretty strong firing rate.

A few milliseconds passed, and the intensity gradient shifted to the next cell. At the first cell, then, the intensity became zero and so the cell slowed up in its firing. A few more milliseconds passed, and then there was a new intensity gradient, this one at about fifty degrees from the horizontal. The cell obediently switched back to a high firing rate—slightly higher than before, in fact. It stayed this way for many thousands of cycles. In fact, it was an unusually constant stimulus, for generally such stimuli changed within a small fraction of a second. But this one just stayed at exactly the same spot for so many cycles that cellular fatigue set in and the cell fired less intensely.

The cell continued to fire sporadically but began to have less and less fuel to supply its energy for firing. Usually fuel was delivered regularly, allowing firing to go ahead without any trouble. But now, for some reason, fuel supplies were diminishing, and, like the clicks of a winding-down music box, each individual pulse threatened to be the very last one, and yet the next always just barely managed to occur.

And then one time it didn't. There wasn't enough energy to make it go one more time. The cell stopped firing altogether.

Scale Five

A ribosome was clicking away, codon by codon, down a very long strand of messenger RNA. This strand was exactly like the other strand that the ribosome had run down a while earlier, but it didn't know that. It simply chugged along, and codon by codon snapped a transfer-RNA molecule into place and fastened an amino acid onto a growing polypeptide chain.

Click-click-click-click . . . Over and over, repetitive work. This ribosome was inside a mitochondrion, inside a retinal cell, inside a—but that is hardly the point. It was doing its job. It had plenty of ATP floating nearby to power the reactions that snapped off the amino acids from the strangely shaped tRNAs. The ATP molecules were small and floated near the ribosome, where they were automatically attracted whenever tRNAs entered a certain channel and were guided into a special narrow chamber deep within the ribosome.

Usually the density of ATP molecules was such that several thousand of them clustered around the ribosome. Right now, however, there were only a couple of thousand, and so the ribosome's progress was slightly retarded. Consequently, this particular copy of this familiar enzyme came off the assembly line in about twice the time it usually took. As the protein grew, new ATPs floated into the vicinity of the ribosome to take the place of ones that had been used. However, there were slightly fewer, even, to replace the used ones, and the ribosome consequently worked a little more slowly. Still, it chugged smoothly down the irregularly winding strand of messenger RNA.

As the ATPs got sparser, the ribosome slowed down. After a while, there were too few ATPs around to make the ribosome really work. Every once in a while it clicked down one more codon, but eventually it stopped.

Scale Six

An electron was circling a phosphorus nucleus, fairly far out and therefore in a fairly classical orbit. It was an easily detachable electron and was attracted by ions of all sorts when they passed nearby. This kind of electron was sure to be pulled away from the phosphorus nucleus within a very short amount of time—at least statistics would have predicted so up till now—for all the other analogous electrons had been snatched away from their nuclei quickly.

But somehow, the proper ions were simply not passing by at the right distance. The phosphorus atom was not getting linked up with the proper partners. So this particular electron, instead of do-si-do-ing its way from one atom to another, continued to cycle rhythmically and periodically around its mother nucleus.

II. EXPANSIO

Scale One

The funeral was Friday morning, and her whole bridge club turned up to mourn the loss of their friend who had been so eager to join them that snowy Wednesday night in Oskaloosa. The minister uttered a moving prayer to the Lord, offering thanks for the time on earth that her soul had enjoyed, comfortingly reminding the gathered sorrowful ones that her soul had gone on to a place of peace and rest and joy. Amen.

Scale Zero

The Earth continued to spin and to revolve about the Sun. It did so many, many times in a row. After a few million such rotations, it was a bit closer in to the Sun than before, although not much. It was a little hotter, on average, partly because it was a little closer and partly because the Sun was burning its fuel differently.

After a while the Sun ballooned and its gases swallowed up the Earth and expanded far beyond it.

By this time the galaxy had rotated six times and was approaching another galaxy. Twenty rotations later, the two galaxies interpenetrated, and for a while they passed through each other like two ghosts or two ripples in water. A few million stars were destroyed, but most were unperturbed. Then after about two more rotations, the two galaxies came apart and went their separate ways.

Kees Boeke was a Dutch schoolteacher with a deep sense of wonder about the universe. In an attempt to share this sense with others, he wrote a classic little book called *Cosmic View: The Universe in Forty Jumps*. The book is based on a simple but powerful idea. It consists of two series of pictures: the first, beginning with a little girl sitting in a chair, zooms out over and over again, first revealing her schoolyard, then her town, then all of Holland, all of Europe, eventually engulfing the earth, the Solar System, the Galaxy, and so on to the murky edge of the known universe. A second series, starting again with the little girl, zooms inward, focusing first on a mosquito sitting on her hand, then on a nearby crystal of salt, a skin cell, a molecule, an atom, and eventually reaching the even murkier domain of elementary particles. Each of the forty jumps either increases or reduces the linear scale by an exact factor of ten, so both progressions are very smooth. The pictures are carefully drawn and clearly annotated, resulting in a book that does a wonderful job of imparting a sense of profound humility and awe for the many-tiered and mysterious universe we inhabit. Boeke's *Cosmic View* inspired a short film bearing the same name, as well as several other books, most notably *Powers of Ten,* by Philip and Phylis Morrison and Charles and Ray Eames.

WHERE'S HOME, COWBOY?

DENNIS OVERBYE

On a clear day, from the grounds of the Jet Propulsion Laboratory you can see the white domes atop Mount Wilson, where astronomers first descried the expansion of the universe. The Laboratory is a 176-acre sprawl of sheds, fences, old missile-range parts, and stone-and-glass high rises bristling with antennae and microwave dishes that hugs the base of the San Gabriel Mountains east of Los Angeles, where Pasadena ranch-houses give way to rocky dry washes and yucca slopes. In the sight of those white domes a mile above you there is a mythic compression of scale. Up there, the ghost of Edwin Hubble, dressed like an English gentleman, pads through dusty Victorian machine shops and past shuttered telescopes, dreaming of pinwheeled galaxies. Down here in the arroyo, pale men in short sleeves and pocket protectors, speaking acronyms as dry as an old saddle, swagger like cowboys to their computer terminals, steer spacecraft around planets, and photograph Martian sunsets. Mount Wilson was once the base camp for the exploration of the universe; for the last thirty years, JPL (owned by the California Institute of Technology) has been the base camp for exploration of the Solar System.

As I drove up the arroyo on a Sunday afternoon in August 1989, past a line of television trailers and fingers of marine mist masquerading as smog, that juxtaposition seemed almost too good to be true. The *Voyager 2* spacecraft was approaching Neptune, currently the planet farthest from the Sun, the last planet in a two-spacecraft, twelve-year, 6-billion-mile odyssey of worlds, moons, rings, and swinging carousels of radiation. Within a week *Voyager 2* would be past the planets, exiting the known Solar System and the domain of JPL and scratching, however feebly, at Hubble's larger universe.

That seemed like a passage fairly begging to be celebrated, and NASA's flacks had certainly succeeded in getting everyone else to agree. Some two thousand journalists were expected to show up for the week-long flyby. Satellite dishes clogged the arroyo. Chuck Berry and Dan Quayle were on the program. It all seemed both miraculous and sad, but then everything about the *Voyager* project has always seemed both miraculous and sad. I was wondering what concatenation of planetary and technological wonders could do justice to the solemnity of this occasion.

I had first driven up this arroyo nearly ten years before, on the eve of Ronald Reagan's election as president—a not-irrelevant milestone, as it turned out—to watch *Voyager 1* go past Saturn. My marriage was falling apart, and the weather was turning gray and gritty in New York; space was the ultimate road movie, and Saturn seemed not too far to flee. As the self-appointed space expert for a glossy new science magazine, I wore my new three-piece wool suit around the lab in eighty-degree heat to meet the head scientists and be told information that would become obsolete the next day. Television monitors occupied every nook and cranny of the lab, showing live pictures from the spacecraft; five minutes after walking in, I, a civilian, knew more about Saturn than the ten generations of astronomers who had followed Galileo's discovery of the rings. Being at JPL was like riding the nose of *Voyager*. Late one night a week later, I found myself far up in the foothills above Pasadena sitting on a floor with a crowd of scientists and writers. Whiskey and other condiments were going around; the air was a haze of cigarette smoke. Jonathan Eberhart, the renowned folksinger and space reporter for *Science News*, was singing space ballads in a deep, ravaged voice. Blasted, floating somewhere between here and *Voyager*, I was falling in love with the planetary encounter. I couldn't believe people actually got paid for doing this. I drove down the mountain at 3:00 A.M., shaking.

I'd been returning to JPL—a cowboy paradise if there ever was one—regularly, sucking stale coffee at midnight while sweating out the

Voyager's dangerous close encounters with rings and moons, sitting in the background while the lab's aggressive young director, Bruce Murray, finagled for a spacecraft flight past Halley's comet and failed, and climbed on giant radio antennae with astronomers who planned to use them to search for extraterrestrial signals. It was there in a crowded pressroom in January 1986 that I spent a gray morning with my colleagues, numbly watching the networks replay the horn-shaped cloud of the *Challenger* explosion on half the screens while on the other half the planet Uranus hung, its rings like a bull's-eye.

I'm a member of what a JPL astronomer my age, a failed astronaut, once described to me as "the cheated generation." We were schooled in the desperate and heady years of *Sputnik,* atomic doom, and the race to land men on the Moon. To us had fallen the solemn and exhilirating task of taking the first steps away from Earth. Our progeny, we grew up believing, would be running all over the Solar System. Every year, armadas of discovery were being sprayed like darts toward the planets—once wandering lights, now places. Questions that had been asked for centuries would be answered in our lifetimes.

But by the time we were adults the space program was already collapsing. "A rat just bit my sister," Gil Scott-Heron sang in 1970, "and Whitey's on the Moon." By the mid-1980s, as the streets filled with the homeless and the hospitals with AIDS patients, it had become harder and harder for even old science fans like me to defend NASA projects that cost as much as the GNP of a small nation. The goal of the agency in the Reagan years seemed to be to make low Earth orbit safe for free enterprise. The space program's most dramatic moments were happening in OMB offices and congressional hearing rooms. By the time of the Neptune encounter, humans hadn't been more than 300 miles from the Earth in seventeen years.

I had written my share of editorials extolling the virtues of space exploration. The main product of the space program, I have always maintained, is not science, nor Velcro, nor military advantage, nor national pride, but consciousness. The Earth is mankind's home, but Earth's home is the universe. After all, weren't our very blood and bones made of star dust, composed of atoms forged in ancient starry dynamos? Was not the wheeling of the Galaxy itself—grand cycles of extinction and mutation —mysteriously inscribed in the evolutionary record of life on Earth? Neptune and Mars and all the rest, their craggy moons, and the comets were kin; surely the stars were as connected to us as our toes. Society was not heeding my wisdom. I had once written that the space program was our monument to the future. But by the end of the decade I no longer believed that there would be humans on Mars or observatories on the far side of the Moon in my lifetime.

In 1936 a Caltech aeronautics professor named Theodore von Karman and a handful of students repaired to a rocky wash in the shadow of Mount Wilson to shoot off liquid-fueled rockets. From such modest scrub-brush beginnings arose a cultural and technological landmark. Von Karman's successors developed tactical nuclear missiles for the army and launched the first American satellite, *Explorer 1,* with one of those missiles in 1958. When NASA was formed, JPL was given the assignment of managing deep-space unmanned missions for the new space agency. The lab's first efforts in that direction were not promising. The failure of a series of Moon probes in the early sixties—of the first six, two fell in the ocean, two missed the Moon altogether, and two hit the Moon but sent back no data—led to congressional hearings. A decade later, however, the lab's *Mariner* spacecraft had spun successfully past Venus, Mercury, and Mars, and the Vikings were being prepared to land and dig on the Red Planet. Today about thirty-five hundred people, mostly engineers, work at JPL—officially employees of Caltech, politically wards of NASA. The official JPL oratory never unbends past the monotone with which Murray, one ex-director, greeted *Voyager's* approach to Neptune: "The success of JPL is due to the excellence of its personnel. We've each had our own experience with that." Most of the affection is reserved for machines.

The twin *Voyager*s were the glory of JPL and of planetary probes, but they were not conceived that way. They were proposed in 1972 as a low-budget alternative after NASA refused to spring for a four-spacecraft fleet that would visit the five outer planets. The key to the plan was a rare alignment of those planets—one that would allow a properly aimed spacecraft to hopscotch from planet to planet using the gravitational field of one planet as a slingshot to get to the next one farther out.

The *Voyagers* launched within days of each other in 1977. It took all of JPL's painfully acquired smarts to surmount a series of crises that ranged from freaked-out computers and stuck camera platforms to the simple cold, dark, and loneliness of deep deep space. The spacecraft survived because they were designed to be flexible, although their intelligence would barely surpass that of a modern digital watch. During the long cruise between Saturn and Uranus, the rocket scientists at JPL—from a distance of 1 billion miles—effectively rebuilt *Voyager 2*'s computer and image-processing systems and redesigned its attitude-control system so that its cameras could take longer-exposure, sharper pictures. The *Voyagers* were unmanned only in the sense that nobody was on board. The crew was in Building 264, eight stories of no-nonsense glass and concrete on the hillside above the cafeteria. In there was one guy, for example, whose job during encounters was to calculate the temperature of *Voyager*'s radio antenna to within half a degree and compute the one frequency on which, because of a failed receiver circuit, *Voyager* could hear and receive commands from Earth. Every time the thrusters fired or the spacecraft rolled, of course, its temperature changed.

Having been steered to a close encounter with Saturn's enigmatic moon Titan, which has an atmosphere thicker than Earth's, *Voyager 1* departed the game at Saturn and headed for open space. Swinging from planet to planet like an acrobat climbing a series of trapezes, *Voyager 2* ventured outward alone, ticking off the planets and moons and possibilities for the configuration of worlds. As the *Voyager* decade went on, the Solar System got weirder and weirder. There were ammonia storms, water volcanoes, braided rings, and the strange moons that made the gaseous giant planets into miniature solar systems in their own right. Among the moons were Easter-egg-colored Europa, grooved Ganymede, and fiery Io at Jupiter, as well as Titan, with its smoggy atmosphere and hints of hydrocarbon slushes and methane lakes. There were moons of molten sulphur, moons of rock-hard ice pitted with craters, and moons coated with some mysterious material blacker than pitch.

Once every planet stop, it seemed, Larry Soderblom, a geologist who was deputy leader of the camera team, would rise up with a heavy cast to his voice and a shake of his head to announce that our geocentric imaginations had failed again. Maybe we would learn by the next time.

The intervals between planetfalls stretched into years. *Voyager* sailed 6 billion black miles, past Jupiter and Saturn, away from John Hinckley and David Chapman, the Iran hostage crisis, the Reagan recession, the advent of AIDS, junk bonds, the stock-market crash, Solidarity, Glasnost and Perestroika, Iran-Contra, goat cheese, and lettuces whose names you couldn't pronounce. The American space program gradually eroded around it until, ten years later, we could see *Voyager* for what it was, a magnificent lone piling marking high tide, standing taller and taller as the water slowly receded around it.

Now Neptune. For the last twenty years, the planetary encounter, a combination of science, circus, and public relations, had been the most romantic experience available on our shrinking planet for cowboy-nerds like myself. Now it was about to become extinct. After this week, except for lonely, tiny Pluto, all the major sights of the Solar System would have been seen. We were cresting the last hill. After this the space program would not end, but it would become science. In the future, if the budgets held up, we would study and study and study these landscapes, covering every tectonic knoll and geyser like soldier ants and inventoring every festive complex hydrocarbon cloud. There were monuments yet to be built that would be inscribed with the names of the great explorers of Mars, but that would be in the next century. For now the circus, if not the science, was over. This was the last chance to see something magic.

I crested the line of television trailers atop Oak Grove Drive and was motioned left by a security guard, past half a mile of chain-link fence to the far parking lot reserved for press. Von Karman auditorium, a low glass-fronted sometime exhibition space named after JPL's founder, was full of reporters lounging armrest to armrest behind long, litter-strewn tables. The walls were banked with strategically placed TVs, on which Neptune was floating like a marble in a bowl. I hadn't been there in three and a half years—not since Uranus and *Challenger*. I looked around for my old friends and was confronted by a vista of bald and white heads, paunches, walkers, wheelchairs. The press had fared worse over the years than the spacecraft. My God, I thought, we've gotten old.

Even in the largest telescopes, such as the ones a mile above us on Mount Wilson, Neptune had never been more than a watery droplet of light 3 billion miles from the sun, a ball of hydrogen and helium with traces of methane, ammonia, and nitrogen and a diameter four times that of Earth. Lately, there had been indications that Neptune was surrounded by narrow, fragmentary rings—so-called *ring arcs*.

What was already clear was that Neptune was one of the most beautiful planets *Voyager* had visited—ocean-blue, laced with bands of high white cirrus-looking clouds. Every photograph of it looked like the Earth as seen from high above the Pacific basin. Prominently placed, like a water stain on the globe, was a huge, featureless dark oval that astronomers suspected was a giant hurricane—but to me looked, its edges wreathed with cirrus, like a big eye. It seemed appropriately symbolic that *Voyager*'s last planet looked superficially like its first planet. It looked so sweet, I had to remind myself that the blue wasn't really warm salt water but cold methane—natural gas. You can't really go home again.

The von Karman era still haunts the hinterlands of the laboratory grounds, where wooden barracks-style buildings predominate and an old army missile stands silent sentry. But out front, the architecture is strictly space age. Just inside the main gate a half-dozen high-rise buildings, angular and hard, their tops bristling with antennae and microwave dishes like the brush cuts of punk teenagers, surround a central mall of gardens and fountains.

When I arrived, workmen were setting up camera platforms on the mall area and hauling in plants for Quayle's appearance. Bulky young men whom I took to be Secret Service were strolling around in suits, looking very intensely at everything and everybody. In the corner of the mall, by the path to the cafeteria, was a black billboard/scoreboard with

—

a map of the Solar System and the *Voyagers'* tracks painted on it. Since I had been coming to JPL, this scoreboard had been giving a running total of the number of days and distances from home of the *Voyagers*. Today, for the first time, I noticed there were new data on the sign. *Magellan,* launched May 4, 1989, was 23,914,404 miles from Earth, on its way to Venus. Moreover, it was fifty days to the scheduled launch of *Galileo* toward Jupiter. I watched as lab visitors posed for pictures in front of the billboard.

The next morning I made the first of many visits to the third floor of Building 264, where, behind the protective bulk of a middle-aged Pinkerton, the hundred or so official *Voyager* scientists were temporarily jammed into a maze of tiny offices. In a corner office, Ed Stone, the *Voyager* project scientist, the man in charge of coordinating the scientific activities of the spacecraft, was munching carrots as he raced through a bag lunch. Stone is short and quick; his hair was brushed straight back over his crown as if he spent his days staring into a jetstream. A Caltech professor and cosmic-ray physicist, he admitted being surprised and a little unsure when he was drafted to lead this planetary adventure back in 1972.

"It's been a lot of fun. I have a whole new vocabulary," he announced briskly. "This probably won't happen to me again. You're project scientist on a mission like this only once—if you're lucky. When we started in nineteen seventy-two, we had no mission to Uranus, we had no trajectory to Uranus. We weren't allowed to consider a Uranus trajectory until nineteen seventy-five." NASA, he said, had finally allowed them to draw up plans for sending *Voyager 2* on to Uranus and Neptune, but with no promises that the plans would ever be executed.

As an intuitive scientist, what did he think would be the most exciting part of the Neptune encounter? "Triton," he snapped, putting down his carrot. "I think we have the prospect of seeing new physical effects. Intuition tells me that that is where the observations are going to stretch us. Triton won't look like any other moon we've seen."

Which made poetic sense. Triton was, after all, the last outpost, the last port of call, the last chance to see something really weird. Telescopic measurements from Earth had revealed frozen methane on its surface, and methane therefore probably also existed in its presumed atmosphere. Geologists were speculating that Triton could have nitrogen oceans as well—if so, they would be the only oceans besides the Earth's in our solar system.

Triton's putative history was bizarre. Viewed from above the Sun's north pole, all the planets in the Solar System revolve counterclockwise around the Sun, and all of the moons in turn revolve counterclockwise around their planets. Except for Triton and Nereid, which travel clockwise and in orbits inclined to Neptune's equator. This means that they are probably captured asteroids. According to calculations published just before the encounter, by going backward around Neptune in an elliptical orbit Triton would have chewed up all of Neptune's regular moons; tides raised in Triton's crust by the bigger planet's gravitational field would have heated it and kept it molten for a billion years. Triton would have been a 2,000-mile-diameter teardrop careening through the skies of Neptune.

I proceeded on my rounds, following the noise and commotion down a hallway thick with people and television monitors to the offices of the imaging team (the largest by far of *Voyager*'s eleven scientific groups) and its boss, Brad Smith. As head of the imaging team, and thus guru and dispenser of the pictures and their interpretation to a visually oriented media, Smith, a University of Arizona astronomer with a crusty manner, is in some ways the point man for the *Voyager* encounters. He is tall and solid, with graying hair, a small mustache, and level blue eyes that occasionally twinkle. Sunburned and taciturn, he has a dry humor sometimes mistaken for arrogance and a smirky smile that gives you the sense that there is a little boy trying to get out of him.

Among *Voyager*'s scientific instruments, the television cameras are first among equals. This confers a similar status, in the eyes of the press, on Smith, who often leaves the maddening impression of giving information grudgingly, especially during early press conferences when there is as yet little to report.

Reporter getting frustrated: "You aren't willing to say, 'There are *x* number of suspected moons?'"

Smith: "I'm willing to say *exactly* that. There are *x* number of suspected moons."

Reporter: "And *x* is?"

Smith (shrugs): "I just lose track. Could be two."

A sign behind Smith's receptionist says, It's Going to Get Worse. Inside his office is a poster that reads: July 20, 1969. Remember Your Dreams *Then,* of *Now?*

Smith was too professional to get excited about the End of *Voyager;* he was worried that it could happen ahead of schedule. He had expressed some concern in the newspapers that debris or undiscovered rings could be lying in wait to smash the hurtling spacecraft when it crossed the ring plane late Thursday night. Dust, he pointed out, had killed all the probes that had come close to Halley's comet in 1986. Passing Saturn in 1980, *Voyager 1* recorded thousands of dust hits from the leftover raw material of the rings. Of all the planets *Voyager 2* had visited, Neptune was the least well charted. With its backward-revolving moons, Neptune was the Wild West of ring systems.

I asked Smith if he really *was* worried, or if it was just something to say to the reporters, to build suspense. His face became grave. "I am extremely scared," he said. "I don't understand why the others aren't." He made an eggbeater motion with his hands. "Ring arcs could be any-where. They could go any which way. We'll breathe a lot easier once we've made the crossing, maybe even sip a little bubbly. If you're not worried, you don't understand the problem."

Triton was waiting on the other side of the ring plane, of course.

In the meantime, no matter who you are, most of what you do during a planetary encounter is watch television. In our experience of space, we've always been watching television, from the days in which we all watched the rockets fall off their launchpads trying to chase *Sputnik* and the wee morning hours we sat up and watched a pair of blurred figures drop onto Moon dust (on the same day that Ted Kennedy drove Mary

Jo Kopechne and Camelot forever into the drink). It's our longest-running show, this collaboration between NASA and the tube. I'm surprised there isn't by now a Space Channel on which the greater and lesser moments are endlessly replayed, a sort of MTV of the sky, an endless witnessing of the Outside. The lesson of space seems to be that almost everything in the universe is somewhere else but doesn't count until we've claimed it with our eyeballs.

A *Voyager* encounter is the ultimate democratic experience. The pictures continuously arriving (after a four-hour journey) from *Voyager*'s two television cameras bounced back and forth between those bristle-topped buildings and then emerged simultaneously on television monitors scattered throughout the environs of JPL and, via NASA, the world. There was one, it seemed, in every office and hallway in Building 264, large clunky sets every few feet throughout the serpentine, sweaty newsroom, a wall of them in the cafeteria.

One screen showed nothing but numbers, real-time *Voyager* data 24 hours a day. Time since launch: 4384 days, 01 hours, 52 minutes 44 seconds; Distance to Neptune: 5,150,175 kilometers; Velocity relative to Neptune: 60,421 kilometers per hour; Distance from Earth: 4,413,556,987 kilometers. The latter number clicked upward by 17 kilometers with every tick of the digital clock. One thing for sure was that at JPL you would never go hungry for numbers.

You couldn't get away from the show even if you wanted to, but who would have wanted to? Those worlds, mute and unblinking, dumb guardians of the truly great beyond, icy cosmic Pillars of Hercules, slowly swelled in the cameras' field of vision. Down in Von Karman Auditorium, the younger, more technically oriented journalists—the so-called press-room imaging team—crowded around each new image, hawkish for details no eye had seen before: Were those craters or clouds? Was it snowing on Triton?

We were scratching for small change, the way pilgrims did at the beginning of every encounter, standing in line for pictures and mysteries that we would forget three days later when *Voyager* was ten times closer to the planet. *Voyager* was normal science speeded up like a time-lapse movie. A week before, the discipline of Neptunology had consisted of a few speculative papers about ring arcs and methane on Triton. A week from now, there would be enough data and dead theories to fill a textbook. In normal science, a major theory might last a year or a century before enough data accumulated to destroy it. In *Voyager*land, whole schools of thought were born, superseded, and forgotten in the span of hours.

The majesty and integrity of science arise in part from its blithe immunity to meaning and significance—a cosmologist I know once left off pondering the universe to write a paper about how his new waterbed worked. It might have been the first planetary encounter of the age or the last—it didn't matter. Like Zen masters, the *Voyager* scientists immersed themselves in the planetary details.

By the middle of the week the ring arcs had become rings. Why was I not surprised? Every planet *Voyager* has visited had had rings. The moons and other mechanisms that were supposed to keep the rings in place hadn't turned up; not one ring system had lived up to theoretical speculation. The Neptunian winds blew too fast and in the wrong direction. The meteorologists, too, had failed at world after world. One afternoon during an informal background briefing, Andrew Ingersoll, a Caltech physicist who was head of the atmospheric part of the imaging team, was asked what we had learned from studying planetary atmospheres, and he gave the same answer he had given during the Uranus encounter: his long, weathered face composed itself in an ironic smile as he said, *"Humility!"*

The 1980s had seen the rise of a theory of physics known as *superstrings,* the most ambitious (and at least for a while, the most successful) step yet in the quest for a mathematical principle that will reduce all the diverse laws and phenomena of nature to an expression that can be printed on a T-shirt. Meanwhile, of course, *Voyager* was sorting primordial ice cubes and discovering that first principles—or even observations on the last planet—were not of much use in predicting the configuration and behavior of the stuff of the cosmos. If *Voyager* was about anything at all, it was about the impotence of theory.

I went to talk to Soderblom about it on the eve of *Voyager*'s closest encounter with Neptune. Soderblom was Mr. Satellites, and I had come to think of him as the resident philosopher on the *Voyager* team as well. Soderblom is pasty and white faced, with mournful eyes and a build that encourages his white shirts to billow outward like tents. His reddish blond hair is cut short and often combed straight over his forehead Julius Caesar style.

Part of the lure of these outer worlds, Soderblom explained, lies in the chemical history of the Solar System. The silicate rocks of which the inner planets are made are a minority constituent of the universe, and all traces of their original form have been erased by weathering. According to theory, they were vastly outweighed in the original presolar nebula— from which the Solar System condensed—by materials like water, methane, ammonia, nitrogen, hydrogen, and helium, but these lighter, more volatile forms of matter were blown out of the inner Solar System by the Sun. A voyage outward from the Sun, to colder and darker realms, is a progressive journey through more exotic and primordial realms of ice. On the Earth, water sometimes freezes; on Mars, carbon dioxide freezes; the Saturnian satellites and rings are compositions in ice. The worlds beyond are primordial ice floes increasingly composed of methane and ammonia.

Triton was the last, coldest, and most distant outpost *Voyager* would visit, and it had captured the attention of the geologists. Triton was shiny, while most of the outerplanet satellites had blackened, which is what happens to methane ice exposed to radiation for a few million years. The implication was that something—weather or eruptions—kept refreshing Triton's surface. This was the geological equivalent of detecting a pulse; Triton was alive. "Maybe we'll see nitrogen volcanoes on Triton," Soderblom mused.

"The thing that impresses me the most," he went on, "is that we keep repeating the same thing. We are always prepared to be faced with the last set of observations. After all, this is the deep freeze of the Solar System. The moons have very tiny mass. We expect too little. Our expectations of what nature can do are too conservative. The only way you can learn is to go out and explore." He looked out the window, as if he could see farther than the San Gabriels bathed in the pink twilight—much farther.

Thursday, encounter day, began on a bad note. Dan Quayle was coming the next morning. Somebody had stolen a box of press badges and run off copies at a local printer, who called the FBI. Faced with the possibility of people walking around with counterfeit badges during the visit of the vice-president, JPL officials decreed that the entire press corps should reapply for new badges. After the morning briefing, we were corralled into sweating queues in the sun to receive from strangers our new, inscrutably misspelled badges, hearing useless admonitions to be at the lab the next morning before 7:00, when the freeways and gates would be closed. Nobody was going home tonight anyway. I marveled at the power, in this democracy, of one man to create so much trouble for people, his alleged friends, just by deciding to drop in on them.

The Sun slowly set on *Voyager*'s last day in the known Solar System. Nobody was going to bed. At about 8:00, JPL time, the spacecraft would cross Neptune's equatorial ring plane, and shortly thereafter it would disappear out of sight as it brushed over the planet's north pole, which is tilted away from the Earth. Then it would reemerge on a course toward Triton, the last outpost, *Voyager*'s last port of call, the last chance for us to see something fantastic, before it headed for the void. Before Soderblom and the rest of us got to see Triton, however, *Voyager* had to cross Neptune's ring plane, making it through what Smith had called "white-knuckle time."

Wearing my new, supposedly counterfeit-proof badge, I went to the cafeteria to eat enchiladas. I've hardly ever eaten in that cafeteria when there weren't television monitors high on the walls, silently—and often, it seemed, unobtrusively—flickering camera images of a distant world. Tonight the room had the speeded-up feel of an airport at night or a newspaper at closing time.

After dinner the newsroom was nearly deserted. Outside, the network trailers hummed with light and energy like jack-o'-lanterns. Smith, Stone, Soderblom, and the others were making the interview rounds.

I wandered into the auditorium. *Voyager*, according to the real-time data, was some 50,000 kilometers from Neptune—the numbers were going down alarmingly fast now, no longer astronomical dreams. Neptune itself was no longer an astronomical dream.

By now, I realized, taking the four-hour light-travel time into account, *Voyager* had already met its fate. It had already been blasted into scrap metal by ice dust, or dazzled by radiation. Or the instrument platform had jammed again, as it had at Saturn eight years before, swiveling at its busiest. We could be watching a ghost, a spaceship that had already died. The knowledge of that ring-plane crossing was on its way to Earth, a little train of radio waves just about now passing the orbit of Uranus. That news was not due home on Earth until after midnight.

Thick black cables crisscrossed the floor of the auditorium, which was lit by floodlights. Television crews were arrayed in a semicircle around the room so that their correspondents could all be shot in front of a full-scale *Voyager* model, which stood against one side of the room.

Perhaps because the constant anthropomorphizing of the plucky little Voyager spacecraft in the press and elsewhere tends to leave the impression of a little flying R2D2, seeing the spacecraft in person was always a shock. It was actually as big as a car. From my perch on the television sidelines, the *Voyager* across the room looked like a giant bug, with its plutonium power-source proboscis and antennae feelers, its heavy tail of cameras, and its radio dish looking like a bony, oversized headdress. On one of its flat sides, gleaming like a gold belt buckle, was the famous *Voyager* record, a compendium of sights and sounds (including a kiss) of the Earth that clever aliens could play back on the enclosed turntable— if they could follow the instructions.

Technically, the radio antenna was *Voyager*'s back, steadily pissing out its 22 watts through the interplanetary black toward Earth, command central, home. You watch the computer animations of encounters and you see the robot arms whirling about precisely as a Balanchine-trained dancer around the periphery of that white circular antenna.

Upon striking metal or anything hard, a dust grain traveling at 60,000 kilometers per hour (40,000 miles an hour, 12 miles a second) literally explodes; a fireball of electricity and energy spreads outward from the microcrater of impact; a blizzard of signals impinges on instruments hungry for detection. I thought of the camera platform rolling now, tucking like a diver, its optics turned protectively backward over its shoulder toward Earth as it, and we, the vicarious riders of the cosmos, plunged toward the plain of dust.

Just before midnight, the screen went black. *Voyager* was through and on the other side of the planet. Suddenly there was nothing to do. In the imaging-team offices, they were drinking champagne. I crawled off to my motel.

I awoke at 3:00 A.M. It was misty outside. As I came up the arroyo, the lights of JPL seemed to float above the landscape like one of Steven Spielberg's creations. In the pressroom the neon lights seemed unnaturally bright. About thirty people were clustered in front of one of the television monitors, whooping and hollering. I couldn't see the screen. At the middle of the knot, a pair of astronomers were commenting on the landscape of Triton.

Another picture came on the screen, accompanied by a long "Oooh!" These were postcards from Triton, and each one looked not just different from, but totally unrelated to the others. Magic postcards. The new image showed a fibrous-looking terrain, which was immediately dubbed "cantaloupe." The next view looked like a field of sand dunes cut by a long Y-shaped rill. The next after that revealed what looked like frozen lakes—craters, said the geologists present, that had filled and remelted and then frozen smooth—lava lakes, only the lava apparently had been water.

Click. "Oooh." The view swung to Triton's limb and the moon's south polar cap. The bottom of Triton looked like every flaky bathroom ceiling that I had ever owned and scraped and painted.

We were witnessing the last moment of planetary discovery, and it felt like an orgy. My detachment had vanished. The early-morning Triton pictures were unexpectedly moving. The landscapes almost glowed, crisp with mystery. Each one felt like a revelation, a jewel. You wanted it never to leave the screen, and you couldn't wait for it to disappear so you could see the next one. Triton was alive; weird things were happening there.

—

It went on and on—magic postcards—as the night imperceptibly lifted. The sky lightened to gray and lightened no more.

And we went out into the mist to face the future and Dan Quayle.

On the mall, chairs and a bandstand had been set up. A royal blue carpet had been rolled down the wide steps of the administration building. A man in a T-shirt was handing out gray and blue balloons. It felt like rain. I was feeling numb and a little let down. My feet and back hurt from standing on hard linoleum in front of a television most of the night. My brain was swamped with contradictory images of Triton.

As head of the National Space Council, a quasi-moribund body resurrected to give the president advice on space matters, Quayle had a semiofficial reason to be here. But his timing made him about as welcome as a bill collector at a wedding. A few weeks earlier, Quayle had committed one of his classic faux pas when, apparently in the grip of a 1950s science-fiction fever, he told CNN that he was excited about sending men to Mars because Mars had canals, which meant it had water. Which meant there was oxygen there. Which meant people could live there.

Quayle strode through the mist, blond and fresh in a gray suit and blinding white shirt. With press-release precision, he sang the glories of *Voyager* and the future of space under Bush and himself. Afterward he stood in front of the *Voyager* model and reminded us of all the great missions to come, missions, like Galileo, Magellan, the Mars Observer, and the space telescope, that sounded oddly familiar, prodigal dreams from the seventies still to be completed in the nineties. NASA had once, in its public-relations wisdom, labeled 1986 the Year of Space Science, when in fact every mission in that year's docket had been postponed from previous years—then, of course, *Challenger* exploded. Some of those missions were still waiting in 1989. Déjà vu all over again. The veep and his muscular entourage were gone as swiftly as greasepaint in the rain.

We were left with marathon press conferences—one day, six different speakers showed the same multicolor diagram of Neptune's lopsided and rakishly tilted magnetic field—and midnight images of snow-cone worlds, melted, pitted, flaking windswept volcanoes, of worlds, the last ones we might ever see. Stone said afterward that Triton would probably be our best look at Pluto for decades. At night, courtesy of Carl Sagan and the Planetary Society, we boogied with Chuck Berry, who sang "Go, go, *Voyager, go,*" to the tune of "Johnny B. Goode" while *Voyager* scientists writhed around him.

In the room where the satellite experts met, a ten-by-fourteen-foot space dominated by a large table and television and computer monitors, where I went for the next few days to watch like a fly on the wall as Soderblom and the others tried to make sense of their strange new world, close-ups of Triton in ever-increasing detail swamped every horizontal surface and then merged into a giant mosaic, computer stitched together. It was in that room and its brethren that I experienced the other magic that *Voyager* had to offer here at the doorstep of the Solar System. Puzzling early one morning over the short, dark streaks that adorn Triton's polar cap, Soderblom was reminded of similarly shaped white streaks, supposedly caused by sulfur dioxide geysers, on Jupiter's moon Io. After a quick chat with his bosses Smith and Stone and a few colleagues, Soderblom marched into the press conference two hours later and put his scientific credibility on the line by announcing that there were probably nitrogen geysers on Triton.

For the next few days, the press wouldn't leave him alone. Soderblom was a man in a crowd, beleaguered but game, as he waited and sifted for the evidence that would prove his hypothesis. From his graduate-school days at Caltech, Soderblom told me, he had always believed that science and the space missions should be open—a sentiment you may not properly appreciate until you've dealt with NASA public relations. Once, I caught his eye in a crowd and his demeanor lifted and he gave me a thumbs-up.

Ten men trooped onto the stage for the final briefing, the fifty-fourth *Voyager* press conference, by Stone's count. Perhaps the last. We were finished, and just in time. The stack of reports, news releases, booklets, handouts, and newspaper clippings on the edge of my desk was a foot high and threatening to topple. The surfaces in my motel room were overrun with papers I wouldn't read. I didn't want to know any more about Neptune or Triton or the evanescent rings. JPL security had slowly eroded; when I went over to the imaging area and got nodded in without showing a badge, there were more reporters, microphones, and cameras standing around than scientists. At the imaging-team party at Ingersoll's palatial San Marino house on Monday night, knots of champagne- and beer-drinking partyers would suddenly find a microphone shoved into the midst of their conversation.

In truth, *Voyager* had been like one long party between the reporters and the scientists, a party to which many of us came in the middle and some didn't stay till the end. Of the eleven original instrument-team leaders, only five were still in the saddle by Neptune—the others had retired or died.

Soderblom stuck to his guns on nitrogen volcanoes. "Just because an idea is crazy," he announced at the final press conference, "doesn't mean it's wrong."

That a man would put his convictions on the line in front of the world seemed as much a miracle to me or as magical as any planetary anomaly, animate or not. One of the gifts of *Voyager* was that it showed what people as well as nature are capable of. I went home satisfied.

A month later, researchers sifting the *Voyager* images found a pair of pictures that showed volcanolike geysers spouting from the surface of Triton. The dark trails of material resembled the plumes from dirty smokestacks. It was suggested that the vents were caused by sunlight heating volatile materials under the surface ice, which then escaped through cracks; similar jets are thought to be responsible for producing the clouds of gas and dust around the heads of comets when they come close to the Sun.

One more trick in nature's repertoire. Should we care, here, as the walls come tumbling down on Earth? It depends, I guess, on where you're coming from. Where's home, cowboy? Once upon a time, it was the circle around the campfire or cave. But as Soderblom says, *home* is an expandable concept. Home becomes the neighborhood, the whole valley, the state, the country, the continent. The Earth. And our concept of kin similarly expands—past family, past clan, past commonality of accent, class, language, race—the farther we travel. On Mars, anybody with a head, anybody *who spoke at all,* would be part of "us." I'd like to believe that the notion and criteria of kinship could continue to grow, transcending even animals and the other manifestations of the genetic code. The Solar System and the life starting to crawl through it are of a chemical piece, derived from the same cloud of star dust billions of years ago, with the same chemical signature as our Solar System has now: 94 percent hydrogen atoms by number; 6% helium atoms, 0.04 percent carbon, 0.008 percent nitrogen, 0.07 percent oxygen. Evolution started with that stardust. Perhaps we should recognize even Triton, made of the same stuff as ourselves, as a member of the family, a wayward distant relative included in the family movies at last.

Why, then, did we like *Voyager* so much? What was *Voyager* for? To a generation raised on tragedy and failure, a generation that saw its leaders fall to bullets and drugs, *Voyager* was a reminder that not everything turns out badly. Not all dreams fail. Sometimes things end happily.

One night that fall I climbed the hill nearby my house to a clearing. Most of the *Voyager* experience was gathered in the early-evening western sky just above the teapot of Sagittarius: Saturn, Uranus, Neptune; the Milky Way, like a chimney of light behind them. Saturn was easy to find, creamy and bright; through binoculars you could almost imagine its rings. Uranus, a bluish green star the same color as in the *Voyager* photos, stood out a little above and to the right of it, almost visible to the naked eye. I strained for Neptune, a dot at best, but never could pick it out among dim stars and nebulae a million times farther away than the little spacecraft—I can't help anthropomorphizing, after all—had yet gone, amid whose chimes of light the swirl of the Galaxy would eventually take it to dance like a dust mote forever.

2 Rms Riv Vu:

On the Search for Habitable Planets

LAURENCE A. MARSCHALL

Rural Delivery, Orrtanna, Pennsylvania, was my address for almost a dozen years. The white clapboard cabin with wide porches on two sides sat in a narrow valley between two channels of Marsh Creek, a rocky stream that tumbles through the foothills of the Appalachians. Just to the northeast was Chamberlain's Hill, separated from the house only by the water and a narrow strip of level forest. To the southwest, hard against the opposite bank, was Mary's Hill. Its summit, no more than a few hundred yards away, was usually lost in branches and foliage, and its shadow, even in summertime, brought an early dusk to the surrounding woods.

I had first been drawn to the valley while hiking among its hemlocks, huge evergreens whose drooping foliage looked almost like hanging shreds of Spanish moss. A cool breeze blew through them, even on August days, and the stream was never silent. Tulip poplars grew there too, though if you knew the species from out on the sunny flats you wouldn't recognize them here. Instead of squat trunks with a heavy crown of spreading branches, these were straight, unbroken shafts of timber, stretching a hundred feet up to the sunlight that seldom reached the valley floor. Farther up the slopes, where the air was dryer and sunlight penetrated more often, stood red oak and beech, with an occasional maple or walnut tree that had escaped the axes and saws of previous settlers. In the springtime, blossoming redbuds and dogwood colored the hillsides.

When I bought it, the house was little more than an abandoned shell, its floor beams punky and its paint mildewed. Little matter; it was the woodland, not the house, that made me stay. Six acres, most of them on the north side of Mary's Hill, came with the deed, but beyond our boundary lines the forest stretched unbroken for half a mile down the valley and for another mile upstream. Where the valley opened up on exposed southern slopes and upland plateaus, there were orchards of apples, peaches, and pears.

Ellen and I would be living there today, but like many of our urban friends who had moved to the country in search of solitude, we found the growth of a family and the demands of work brought back a longing for the company of others. Woodsy life was peaceful, but we missed idle talk with neighbors and the sound of an occasional car passing by. There came a time, when our son was born and our daughter about to enter school, that we thought of the choices that lay ahead. Since we now needed a new bedroom, we could build onto our house—which was already crowding the stream on both sides—or we could move. In town the schools were better and there were sidewalks for playing hopscotch. Our children's playmates would be within walking distance. We would not have to spend so much time driving to work, to the store, or to the baby-sitter. I would no longer wake in a sweat at two A.M., worrying whether the pump would fail, the creek rise, or the composting toilet fail to digest.

One day while driving to work, I saw a For Sale sign on a lawn near the campus. Ellen and I agreed: the building was a bit too large, and the carpeting (which we could not afford to replace) was a hideous purple, but it had character—and two extra bedrooms. And so we moved, to a shake-covered house with a spruce tree beside it, and a large, sunlit backyard. Almost a decade has passed, and though from time to time I miss the brooding hemlocks, the smell of mossy earth, and the sound of the moving water, I know that the big house with the ugly carpet is our home, and someone else lives by the rushing stream.

Stories like this are repeated every day, for we are a mobile species. We change homes to find a job, to escape oppression, or to get out of a bad economic situation. We move because we have to or because we want to, sometimes simply to try out a different way of life. Yet whether we move 10 miles or halfway around the world, we really haven't traveled very far. Culture may vary from place to place, but in a physical sense the world alters very little, for ours is a small planet with a limited range of conditions in which life can flourish.

Looking uphill from my old home, I could see a part of this range of variations. The slopes of Mary's Hill held several microclimates, cold and damp at the bottom, sunny and dry at the top. Change was gradual, of course; there were no sharp dividing lines between the domains of different types of vegetation, and it felt like familiar ground whether one was standing at the hilltop or by the stream bank near my house. There are spots on Earth where the contrasts are more marked: Kilamanjaro's snowy peak is only a few miles from equatorial savanna, for instance, and one can drop a stone from the forested rim of the Grand Canyon to the desert below. But though they are sometimes striking, the distinctions between desert and forest, and between tundra and jungle, are not as extreme as one might imagine. A few score degrees Celsius in temperature, a few dozen inches of rainfall a year are all that separate one place on Earth from another.

Life on Earth has been as successful as it has, I am certain, because of the restricted climate in which it finds itself. Wherever we go, there are life forms that, while adapted to one place, do quite well somewhere else. My daughter found a prickly-pear cactus one day growing on a railroad right-of-way near our house, 2,000 miles from its relatives in Arizona. Shrubs from China and trees from Europe dot our neighborhood, having colonized new territory right along with the humans that brought them. And even at the extremes of climate, life seems recognizably terrestrial. The microfauna that inhabit hot springs and the tubeworms from marine trenches function in ways we can understand; though they seem alien at first, they have kin in other, less exotic, terrains. Our biosphere, the habitable zone of our planet, is no more than 10 miles from ocean bottom to mountaintop, as thin as the skin on an apple when compared to the 8,000-mile diameter of Planet Earth. In this restricted volume, it is no wonder that life has followed a convergent path.

Leave the confines of our planet, however, and the universe becomes decidedly less hospitable. After three decades of planetary exploration, we have found not a single other place on which we would feel at home. The planets closer to the Sun, Mercury and Venus, are too hot, while distant Pluto is too cold. Jupiter, Saturn, Uranus, and Neptune may have no solid surfaces to speak of, not to mention noxious atmospheres of hydrogenic molecules. Temperatures at the equator of Mars, the place most like ours, may sometimes be bearable, but the planet is too arid overall. And while a few astronomers think there may be temperate seas beneath the surface ice of Europa (one of Jupiter's larger moons) or under the perpetual clouds of Saturn's largest satellite, Titan, these seem to me to be last-ditch efforts to find a habitable place in our own solar system. The habitable zone around our Sun includes only one planet, and that is our Earth.

As a species, human beings are far more isolated in the universe than my wife and I were deep in the Pennsylvania woods. Nowhere does there seem to be another place like ours, another planet we could call home. Life on Earth belongs to no biological neighborhood, nor can we locate ourselves within any cosmic biosystem. The biosphere resembles a terrarium, a crystal globe sealed off and floating in space—and for all we know, alone.

Over the last few decades there has been a growing realization of just how unique our situation is. *Spaceship Earth* has become a convenient catchphrase to describe it; and even politicians use the word *biosphere* as freely as ecologists. But I am not certain that the extent of our specialness has really established itself in the collective psyche. It does not sit well; we find it hard to accept the notion that life can be so localized, such a small part of such a large universe.

Somewhere out there, most of us believe, are planets like ours. Somewhere out there is life like ours. Somewhere—and one need not be a UFO fanatic to believe it—there are intelligent beings. In time, perhaps, humanity may even travel to the stars, visit other planets, colonize a place outside our crystal globe. We will shake the hands of our fellow travelers in space and time.

As an astronomer, I share this belief, but I'm aware of how close it comes to being an article of faith. In principle, planets like ours may abound in the universe, for I believe the natural processes that formed our planet were not that special. But as a scientist I know I must remain unconvinced until I see the evidence, and to date there is none. Not a single habitable planet has been found anywhere besides our Earth, neither in the Solar System nor circling any other Sun-like star in the universe.

Does that seem surprising? After all, there are plenty of stars to choose from. The Sun is located in a rather typical galaxy, the Milky Way, a huge spiral swarm containing as many as a hundred billion stars similar to our Sun in size and luminosity. Since the planets of our solar system formed from debris left over when the Sun condensed, it stands to reason that what happened to our Sun probably happened a hundred billion times over in our galaxy. By this reckoning there could be billions, perhaps hundreds of billions, of solar systems in the Milky Way. Surely some of those planets would be neither too hot nor too cold, too gaseous nor too dry to nurture some sort of fertile biosystem. And yet we know of none.

Given this sobering state of affairs, it is some consolation to know that our apparent isolation may be due not to an absence of habitable planets but rather to the difficulty of discovering them. Even the nearest star is so far away that a planet near it could be detected only by pushing astronomical instruments to the very limits of their capability.

Here's the problem: to see a planet, you must be able to distinguish it from the star nearby. But it is impossible to see sufficiently fine detail through the Earth's turbulent atmosphere. On even the clearest of days there are swirls and eddies in the air that scramble light rays from celestial sources. This is the same effect that causes stars to twinkle and makes distant objects viewed over a stretch of summer highway look as if they were painted on fluttering silk.

If we were observing our own solar system from a few light-years away, Earth and Sun would be subsumed into a single featureless blob of light. No amount of magnification would separate the two. A single shimmering blob, in fact, is all we see when we look at any star under the high magnification of a telescope, the size and steadiness of the blob depending on just how stable the atmosphere is on a particular night. The difficulty is compounded by the fact that planets, which shine only by reflected light, are much fainter than the stars that illuminate them. Under the steadiest atmospheric conditions here on Earth, the aureole of glare from a star still swamps the glimmer of its circling planets.

Astronomers had hopes that the Hubble Space Telescope, which observes deep space from above most of Earth's turbulent atmosphere, would be able to resolve these problems. If its mirror had been perfect, the images of the stars it saw would have been steady pinpricks of light ten to fifty times sharper than the best views from the ground. Had it worked properly, Hubble might just barely have been able to tease out the images of planets from a few nearby stars. Optimism was dashed, however, by the well-known optical imperfections in the telescope mirror, which became apparent soon after launching. As seen through the Hubble telescope, each star image is surrounded by a spidery halo of light, which makes planetary discovery with the instrument a hopeless task, at least until corrective optics are installed in the mid-1990s.

If planets are discovered in the near future, however, it may be by other means than simply taking a picture of them. For several decades astronomers have been looking for planets by attempting to measure the effects planets have on their parent stars. This method has the advantage that one doesn't even have to see the planet to detect its presence. What one does, rather, is look at the parent star to see if it moves around in response to the gravitational pull of its planets.

The principle behind this method is as old as the theory of gravitation itself. We know that planets orbit our Sun because they are pulled toward it by its gravitational pull. Each planet, in return, exerts an equal and opposite force on the Sun, pulling the Sun toward itself as it orbits. Because the Sun is so much more massive than any planet, it responds to this force by moving only a small amount, but it in effect orbits, just as the planet does. Planet and Sun, in fact, execute a dance around a point that lies between them called the *barycenter* of the Solar System, like two square dancers swinging each other in place.

—

Thus, even if we cannot see a planet, we can watch the star's part of the dance, noting its position carefully as it sways back and forth in response to the tug of its companion. The shift is small—our Sun, for instance, shifts back and forth a distance about equal to its diameter in response to the pull of the largest planet in the Solar System, Jupiter. But by measuring this motion one should be able to detect the presence of planets near stars even when the planets' images can't be seen directly. One can even get an idea of the size of a planet and its orbit from the size and shape of its parent's wavering path.

Although in principle the wobble of a star can be a sensitive test of the presence of planets, the results to date have been far from conclusive. Emblematic of the difficulty is the work of Peter van de Kamp, now a professor emeritus at Swarthmore College, who pioneered the search for planets. Van de Kamp, a lean, white-haired gentleman who speaks with a measured Dutch accent, is an avid observer and an equally avid devotee of Charlie Chaplin. At one point in his career he would show the campus a comedy film from his own collection each week, accompanying it live on the piano.

Astronomers tend have mixed emotions about van de Kamp. They respect him as a meticulous astronomer, well versed in the classic techniques of measuring star positions from photographs. But they have not been able to accept his conclusions. A number of the nearest stars, he has claimed, show suspicious wobbles, small in size but definitely measurable. Most obvious is a faint red star known as Barnard's star whose motion, according to van de Kamp, reveals the presence of at least one orbiting planet, about as massive as Jupiter.

Van de Kamp's professional contemporaries have come to view these claims as unsubstantiated. The supposed orbital motions of the stars, they argue, are smaller than one could hope to measure using Swarthmore's telescope and current measuring technology. Any apparent wobbling of stars is simply the result of the inherent imprecision of the measurement. Van de Kamp was simply trying to push his instrument too far, using a foot rule, in effect, to measure the size of a speck of dust. This has not sat well with van de Kamp, who has seemed to take honest criticism as an attack on his integrity. I recall him rising angrily at professional meetings to defend the reality of the extrasolar planets. Despite the controversy, van de Kamp got a great deal of public attention for the encouraging news that our solar system is not alone. To this day astronomy textbooks still cite Barnard's star as being a likely candidate for a nearby planetary system.

Perhaps it is, but the majority of astronomers still find van de Kamp's evidence uncompelling. The problem is the Earth's atmosphere. Because the images of stars are so far, and because the expected wobbles are so small, it is difficult to measure star positions with sufficient precision to uncover any planet-induced motions. Even improved recording and measuring techniques have not improved things much. Space-borne telescopes like the Hubble may in time be able to tell us whether van de Kamp was right.

In the meantime, there is a much more sensitive alternative technique that measures the motions of stars toward or away from us. As Jupiter orbits around the Sun with a period of 12 years, the Sun moves in response. If we were to watch this from afar, and if we were looking along the plane of the orbit, we would see the Sun swing toward us with a speed of as much as 30 miles per hour during half of Jupiter's orbit, and we would see the Sun moving away from us during the other half. The motion of the Sun could be detected, in fact, even if Jupiter could not be seen at all. By taking spectra of a star, and by carefully measuring the wavelengths of light in each spectrum, astronomers routinely measure such motions of stars. They make use of a phenomenon called the *Doppler shift:* the wavelengths of light received from a moving source are shortened when the source is approaching and lengthened when the source is moving away. The radar guns used by highway patrolmen employ the Doppler shift, automatically comparing the wavelength of a radio wave sent out to the wavelength received from a moving vehicle.

Until recently, Doppler measurements of starlight were limited to precisions of about 2,000 miles per hour, not good enough for detecting planets. Techniques have been improving substantially in recent years, and there are several projects under way that should be able to detect stellar motions due to planets of the size of Jupiter or even smaller. Yet even though the instrumentation for planetary searches is available, astronomers still face a formidable hurdle: the limitation of time. To reach a firm identification, an astronomer must watch a star for several orbits of its planet, not just a few days or months. It's not sufficient just to show that a star is changing its velocity. One must also confirm that the velocity is changing in a fashion consistent with the presence of an unseen attractor. To identify a Jupiter around another star, consequently, would take over a dozen years of repeated observation; to find a Saturn (which orbits farther and more slowly) would take a lifetime.

Scientific projects of this duration are difficult to maintain, in part because lives are short and in part because funding agencies are impatient for results. Bruce Campbell, a young Canadian astronomer who developed an exquisitely sensitive device for measuring stellar velocities, retired from the field in 1991 after a decade of scrambling for support. Though Campbell's instrument, a spectrograph capable of timing stellar motions as slow as a few meters per second (the speed of a brisk walk), yielded a steady stream of data, there were no headline-grabbing discoveries. Campbell found it nearly impossible to obtain a equally steady stream of financial support, and he finally opted for a dependable paycheck and a day job. He left behind a wealth of data that suggest that several nearby stars may indeed be feeling the tug of nearby planets. A number of other researchers continue along the same lines, but it will be some time before definitive results are in.

Progress in the search for habitable planets goes on at a slow pace, largely unnoticed by the public media. So fondly do we long for other worlds, however, that anything that can be construed as a planetary discovery, no matter how bizarre, finds a ready audience. In July of 1991, news articles heralded the detection of a planet that shouldn't have been where it was. The new planet, about the size of Jupiter, was reported to be orbiting a remote pulsar, a spinning star that sends us pulses of radiation like the beams of a lighthouse. It had been detected by British astronomer Andrew G. Lyne and his co-workers, who had not been looking for planets, but simply measuring the regularity of pulsar spin rates. As they graphed their data, they noted regular changes in the time of arrival of pulses from the star that could only be due to a regular motion of the distant star in response to the pull of an unseen planet.

If Lyne had been correct, this would have been an odd state of affairs. Neutron stars are dead cinders of stars, all that is left after a star explodes. Such explosions—called *supernovae*—should wipe out any planets orbiting around them. It was possible, some astronomers suggested, that the planet had formed *after* the explosion, from outgoing shreds of the dying star, but the notion was so bizarre that even nonastronomers were skeptical. "Of course there is the possibility," wrote a perceptive reporter for the *New York Times* on July 28, "that there is no planet after all."

The *Times* was correct. There was no planet. Six months after the announcement of the new planet, Andrew Lyne rose to address the American Astronomical Society. To a thousand stunned astronomers—who had expected him to describe the circumstances of his remarkable discovery—he announced that he had been misled. The apparent changes in pulse times, which he had attributed to a planet, were due to the way Lyne and his co-workers had corrected their clocks to account for the Earth's motion around the Sun. The consequent hubbub at the meeting overshadowed the announcement by an independent group of astronomers of a probable pair of planets around another distant pulsar. They had carefully checked their work for errors of the sort that had tripped up Lyne and were confident that they had made no such mistakes.

This latest planetary detection looks to me to be a sound one, but it is no cause for optimism in the search for habitable planets. There would be no streams or sunlit slopes on planets such as these, for their parents are dead stars packed with matter shrunken to unbelievable density. Seen from its orbiting planet, the central star in one of these systems would be a glowing sphere no bigger than a terrestrial mountain, emitting intense beams of X rays and radio waves as surrounding matter winds through the star's magnetic field and crashes into its surface. Such planets would be the ultimate nuclear disaster zones, scorched, barren, and unlivable.

I do not doubt that someday we will know of other habitable planets, and it looks to me as if that day might come within the next decade or two. The difficulty of the task, however, should make us better appreciate the need to husband what we have close at hand. Good habitation is not easy to come by.

When I was packing up to move from our home in Orrtanna, I came across a poem I had clipped, "Cascadilla Falls," written by poet A. R. Ammons, who was teaching at Cornell in the 1960s. The words seemed just right for someone leaving the banks of Marsh Creek. In the poem, Ammons picks up a stone and throws it into a stream near campus, thinking

> all its motions into it,
> the 800 mph earth spin,
> the 190-million-mile yearly displacement around the sun,
> the overriding
> grand
> haul
> of the galaxy . . .

As the stone settles into the rushing waters, Ammons has a moment of fleeting despair for the smallness of the place in which he finds himself. Leaving my old home, I felt the same way.

> shelterless,
> I turned
> to the sky and stood still:
> oh
> I do
> not know where I am going
> that I can live my life
> by this single creek.

Ammons's words return to me with renewed resonance as I think about places so far from us we can only guess at their existence. Looking up at the sky, miles from the single creek I used to call my home, I know that I live my life there still.

—

THE DISCOVERY

ROBERT M. HAZEN

Scientific discoveries are few and far between in most scientists' careers. Most scientific research proceeds at a halting pace. Data accumulate gradually by tedious, repetitive experiments and painstaking analysis. But 1988 at the Carnegie Institution of Washington's Geophysical Laboratory was a most unusual year. In January and February we identified three new superconductors, all of economic importance. During the summer months we found three new minerals, each with a beautiful, previously unknown atomic structure. But perhaps the most memorable discovery was saved for last.

The crucial experiment started routinely enough at about noon on Thursday, December 22. That morning I had played hooky with my wife, Margee. We arrived early at a blissfully deserted shopping mall to complete our Christmas chores. It was almost 11:00 A.M. when I drove up the winding, tree-lined driveway to the Geophysical Laboratory's hilltop site in northwest Washington, D.C. My research partner, Larry Finger, was waiting. We share the X-ray crystallography lab, where we study the atomic arrangements of interesting minerals and other crystals.

The concept of X-ray crystallography is simple. Shoot a focused beam of X rays at a crystal. Some of the X rays will scatter off planes of atoms in certain fixed directions, and by measuring the directions and intensities of those reflected X-ray beams, or "peaks," we can deduce the atom positions. In practice, X-ray crystallography requires fancy, expensive machines called X-ray diffractometers that orient crystals and collect data automatically. Our lab has four of the beasts, each with precisely machined geared circles and jutting arms, each with its own distinctive personality.

Larry and I hoped to keep the diffractometers grinding away over the holidays studying a synthetic crystal of magnesium silicate, denoted $MgSiO_3$ because of the 1:1:3 ratio of its elements—magnesium, silicon, and oxygen. One form of magnesium silicate, the hard, usually dark mineral pyroxene, adopts one of the most common atomic structures in nature. In its transparent pale green variety it sometimes serves as a semiprecious gemstone. But we were exploring the atomic structure of a different form of $MgSiO_3$. As you squeeze pyroxene at above approximately 200,000 times room pressure, the atoms get uncomfortable and shuffle around to form a denser structure—ilmenite. Finally, at about 300,000 times atmospheric pressure, the $MgSiO_3$ in its ilmenite form transforms again as the atoms shift into yet another arrangement, called the perovskite structure. Those "transformations" cause big increases in density, which account for some of the seismic blips deep in the earth. The earth is in effect an onion with thin layers of denser and denser stuff the deeper you go. Scientists believe that the ilmenite form of $MgSiO_3$ contributes to one of those layers.

The crystal we were to study looked quite ordinary. It was a colorless, clear speck about 0.01 of an inch across. Japanese researcher Eiji Ito created the specimen in a giant high-pressure, high-temperature device used to synthesize diamonds and the like. Everything in that Japanese synthetic run was supposed to be $MgSiO_3$ ilmenite. But high pressure is tricky—you can never be sure what a million pounds of force at temperatures of thousands of degrees might do to magnesium and silicon.

Larry and I mounted the crystal on our new Rigaku diffractometer. The automated marvel did most of the work for us. We went off to lunch, and by 3:00 P.M. the Rigaku had located peaks and revealed the dimensions of the regular atomic lattice.

Something was clearly wrong. Crystals can be described in terms of regularly repeating boxlike units called unit cells, which stack together to form the crystal. The size and shape of the unit cell is a kind of crystallographic fingerprint. Magnesium silicate ilmenite should have unit-cell dimensions of about 4.72 by 4.72 by 13.5 angstroms (an angstrom is a minute distance, just one ten-billionth of a meter), but our crystal was different. Its unit cell was 4.75 by 4.75 by 12.95 angstroms: close, but entirely different. For one thing, it was 4% denser—a huge difference when talking about the earth. We went over the possibilities. The similarity of the two unit cells was too close to be a coincidence—there was some relationship between ilmenite and our stuff. Could the Japanese scientists who first described the material have made a huge crystallographic boo-boo? Was it a typographical error in the original paper that described the ilmenite? No, others had confirmed the ilmenite numbers and density. We spent the next two hours trying to check out all the possibilities, looking at peak shapes and measuring intensities. I began to suspect the only remaining logical conclusion—we had discovered a new high-pressure variant of $MgSiO_3$, a new atomic structure. After all, our crystal was denser and its close dimensional relationship to the ilmenite phase was obvious. Larry started an overnight data collection that would allow us to refine the structure in the morning.

Larry had to leave at 5:30 P.M., but I stayed around the lab and resolved to talk to Rus Hemley, a lab colleague and mineral physicist who had done a lot of thinking about the high-pressure transformations of $MgSiO_3$. I gave him the facts. We had a crystal of $MgSiO_3$ that was supposedly synthesized in the ilmenite stability range. The unit cell was very close to ilmenite's, but our phase was 4% too dense, with a slightly longer *a*-axis and a much shorter *c*-axis. It only took Rus a second to respond.

"Have you checked out the lithium-niobate structure?"

His question rang only the faintest of bells. As I mumbled something about not having had time to check that possibility he retrieved a full folder neatly labeled ilmenite/$LiNbO_3$ and handed it to me. The dark green, well-worn file contained a dozen dog-eared reprints and preprints on the rare structure of lithium niobate as well as recently documented high-pressure transitions from the ilmenite structure to this $LiNbO_3$ type. The density change was 4 percent. The *a*-axis got longer. The *c*-axis got much shorter. And in one of the preprints the authors (one of whom was our lab's director, Charlie Prewitt) suggested that $MgSiO_3$ in the ilmenite structure might transform to $MgSiO_3$ in the lithium niobate structure at high pressure.

We had made a discovery of tremendous importance. Deep within the earth is a 100-kilometer-thick zone containing ilmenite-type $MgSiO_3$. Mineral physicists claim that at a depth of 670 kilometers ilmenite transforms to the much denser perovskite form. The trouble is that seismologists, who first identified the abrupt 670-kilometer discontinuity on the basis of reflected seismic waves, say the density change is all wrong—mineral physicists, they complain, see much too large a density change in their experiments. Now we had the solution to this conflict. Instead of transforming from ilmenite to perovskite all at once, the crystal undergoes an obvious structural transition from ilmenite to lithium niobate to perovskite, a much gentler change that takes place over a thicker layer.

Our work had taken on a new urgency. This was big news—the kind that makes a lead article in prestigious periodicals like *Science* or *Nature*. The familiar tingling feeling of being on the track of something big crept up.

I called Larry at his home and we agreed to meet early the next morning. I also reached Charlie Prewitt and told him of the discovery that matched his predictions. Home, finally, at 8:00 P.M., I remained in a thoroughly agitated and exhilarated state. Thank goodness the Christmas errands were out of the way. Margee and the kids listened patiently as I described the data and their significance. Then I phoned Lab colleague Dave Mao, who had first obtained the crystals from Japan and was supervising their distribution. He, too, was excited and immediately grasped the importance of the findings. I spent the evening plotting strategy, particularly the manuscript preparation that would have to begin immediately the next morning.

Sleep was a lost cause. I was up at 5:00 A.M. Margee suffered a largely sleepless night thanks to my incessant jostling, but she got up to cook a sustaining egg breakfast. We were extra-quiet so as not to disturb the children (who had been sleeping poorly in anticipation of Christmas) and the dog (who, once aroused, would wake up the children). I sneaked out with a quick good-bye kiss and got to the lab before 6:00 A.M.

The diffractometer had completed its task, so I gave it another crystallographic chore and headed to my word processor. The prose flowed smoothly, for this was an elegant experiment with unambiguous results and profound conclusions. There was a sense of urgency in my writing, too, for the lab's Christmas party was to begin at 1:00 P.M., and after that there would be no hope of soliciting secretarial help. Larry also arrived early, as promised, and he went straight to work analyzing the diffractometer output. By 9:30 A.M. we were ready to refine the crystal stucture. It was an easy task because previous authors had described the lithium niobate structure clearly. All we had to do was substitute magnesium for lithium and silicon for niobium in the computer description. And bingo—on the first try it refined to a disagreement between our model and our data of less than 10 percent, a good, low number indicating that we were clearly on the right track.

Two hours later we were still on the right track, but things weren't getting any better. We could locate atoms well enough, and all the presumed lithium niobate atom positions were properly occupied by atoms. The trouble was that the atoms didn't look exactly like silicon or magnesium. We assumed that our problems were all the result of twinning —a nasty habit crystals have of not being quite regular. Larry had just written a sophisticated computer program that analyzed for twinning, but it took time to run. The minutes ticked by: 11:45, 12:00, 12:20, 12:40. The Christmas party was approaching, and still no solution. Some of our models matched the data almost perfectly, but the atomic vibrations ended up being much too small, and sometimes even negative—an impossible result. What was going on?

We quit for a while to enjoy the traditional lab Christmas-party fare—a fresh roast-beef sandwich, garden salad, apple pie, and beer. Larry and I were the center of attention, surrounded by other scientists eager to hear the latest details of the find. Charlie Prewitt was beaming, having helped to predict the transition before its discovery. Drs. Liang-chen

Chen, Ming Sheng Peng, and Jinfu Shu, postdoctoral fellows from the People's Republic of China, kept their cameras busy as they took dozens of pictures of the party; Larry and I were asked to pose with each for commemorative portraits. It was quite a celebration, and after a couple of beers we really got into the swing of the celebrity status that any significant scientific advance affords. That we had not yet solved the structure was a minor fact, easily forgotten.

It remained only to tidy up details of the recalcitrant structure refinement. We headed back to the computer terminal and were working on another twin model when Dave Mao walked into the office and sat down. He had been following our work with keen interest all day and had obviously been doing a lot of thinking about the problem. You can never be sure what Dave is going to say, but his remarks are always well considered and deserving of respect. His question was devastating.

"Have you considered corundum?"

Do you know the feeling when you've done something incredibly silly or profoundly dumb and it just sort of slowly creeps up on you what an idiot you've been? Corundum's structure—Al_2O_3—is closely related to that of ilmenite. The crystal structures are actually almost identical, because the atoms are arranged the same way. The unit cell of corundum is 4 percent smaller than ilmenite's. It has a longer a- and shorter c-axis. The only structural difference is that 2 aluminum atoms (with 13 electrons each) take the place of a magnesium and a silicon (with 12 and 14 electrons, respectively). Corundum is everywhere in our lab. We use it as a high-pressure calibrant, as an abrasive, and as an X-ray standard. And the Japanese use corundum to pack high-pressure experiments. There were a hundred ways that a tiny corundum crystal might have contaminated the capsule of $MgSiO_3$. We had been so intent on finding the magnesium silicate that we ignored the simple and obvious alternative. All that fuss over a piece of grit! What could we say or do?

I guess everyone reacts in his or her own way. Larry said nothing. He turned to the computer for reassurance, his jaw set, his expression unrevealing. Quickly he attacked the keyboard, magically transforming silicon and magnesium into 2 aluminums. For that was all we really had to do. The refinement converged quickly and smoothly, a perfect agreement with the common corundum structure.

At first I refused to believe it. Surely the lithium niobate story was true. It seemed so right from an esthetic point of view. But at the same time I knew corundum was the answer. Scientists are taught to look for the unexpected but always to accept the simplest answer. And I could only laugh that we had, in our unthinking, delirious, greedy rush for science headlines, missed the most obvious and trivial reason for the anomalous data. Our crowning achievement of 1988 had turned to dust. Even worse, we had trumpeted our "discovery" to our colleagues, before all the data were in.

By 5:00 P.M. the Geophysical Laboratory was deserted. Larry had left for home, looking forward to a week in the Grand Cayman islands with his wife and daughter. He would soon forget the embarrassment. Dave Mao, Charlie Prewitt, and Rus Hemley had also left for their homes. They had said little upon hearing the revised news that we had "discovered" corundum. Perhaps they were chuckling inside, thinking that in a few months our folly would make for a funny story over a beer at some professional meeting. But they were also sensitive enough to know that the jokes should wait. I alone remained at the lab, determined to make one final test of that deceptive crystal. Chemical analysis would tell for sure if we had aluminum oxide or magnesium silicate. Of course, I knew the answer before it appeared on the microprobe output. I called home, gave Margee a brief synopsis of the day's depressing results, and headed home for the holidays.

I got home at 7:30 P.M., exhausted and discouraged. The street was quiet and the house dark. I trudged up the front steps and was fumbling for my front-door key when the mail slot popped open, and a small hand appeared, silhouetted in a rectangle of light.

"Dad's home!" Elizabeth called out.

Almost at once the door swung open, and Margee was there with a hug and a kiss. Ben yelled out "Hi!" from the TV room, and our dog,

Yipko, ran her usual excited circles, alternately jumping on my leg and tugging at my shoelaces. They all seemed so happy to see me—they didn't care at all that I had failed.

And it struck me that in the whole scheme of things, it really hadn't been such a bad day after all. Science is a great way to spend a life, and there will doubtless be other good years with other exciting finds. But that is only part of life's equation.

It's not often that you make such an important discovery.

CONTRIBUTORS TO
MYSTERIES OF
LIFE AND THE UNIVERSE

DIANE ACKERMAN, a poet and an essayist, is the author of ten books, including *A Natural History of the Senses*, *The Moon by Whale Light*, and *Jaguar of Sweet Laughter: New and Selected Poems*. She is a staff writer for the *New Yorker*.

ANTHONY AVENI is the Russell B. Colgate professor of astronomy and anthropology at Colgate University in New York. He is the author of *Empires of Time: Calendars, Clocks, and Cultures* and *Conversing with the Planets*.

MICHAEL H. BROWN, an investigative journalist, is the author of six books, including *The Search for Eve*, *The Toxic Cloud*, and *Marked to Die*. His work has appeared in *Rolling Stone*, the *New York Times Magazine*, the *Atlantic Monthly*, *Discover*, *Audubon*, and other publications. He has served as a contributing editor for *Science Digest*.

JAMES O. FARLOW is a professor of geology at the Fort Wayne campus of Indiana University. He is the author of numerous technical and popular articles about paleontology and dinosaurs. He was also the editor of a collection of technical papers, *Paleobiology of the Dinosaurs*, and has written a children's book, *On the Tracks of Dinosaurs*. Dr. Farlow has appeared on "Nova" and other television programs dealing with dinosaurs.

TIMOTHY FERRIS is the author of *Coming of Age in the Milky Way* and five other books on astronomy, physics, history, and philosophy. He is professor of journalism at the University of California.

DAVID H. FREEDMAN is a contributing editor for *Discover* magazine. He also writes for *Science* magazine and the *Boston Globe,* and is completing a book on artificial intelligence.

MARTIN GARDNER has written for many publications, including *Scientific American,* where he was a columnist. He is the author of numerous books combining math, science, philosophy, and literature, such as *The Whys of a Philosophical Scrivener* and *Penrose Tiles to Trapdoor Ciphers.* Among his most recent books are *The New Ambidextrous Universe* and *More Annotated Alice.*

JAMES GORMAN is the author of *The Man with No Endorphins, The Total Penguin,* and *First Aid for Hypochondriacs,* among other books. His work has appeared in a number of national magazines, including the *New Yorker,* the *Atlantic Monthly,* and the *New York Times.*

BRUCE GREGORY is an associate director of the Harvard–Smithsonian Center for Astrophysics. He is the author of *Inventing Reality: Physics as Language.*

ROBERT M. HAZEN is staff scientist at the Geophysical Laboratory, a nonprofit research center supported by the Carnegie Institution of Washington. Hazen is author of *The Breakthrough: The Race for the Superconductor,* and coauthor, with James Trefil, of *Science Matters: Achieving Scientific Literacy.* He is a Clarence Robinson Professor of Earth Sciences at George Mason University.

DOUGLAS R. HOFSTADTER is a professor of cognitive science and computer science at Indiana University, where he studies error-making, analogy-making, and creativity and directs the Center for Research on Concepts and Cognition. He is the author of several books, including *Metamagical Themas* (a collection of columns he wrote for *Scientific American*) and *Gödel, Escher, Bach,* which won the 1980 Pulitzer Prize.

LAWRENCE E. JOSEPH, a free-lance writer from Brooklyn, has written for a variety of national publications, most frequently the *New York Times Magazine.* He is author of *Gaia: The Growth of an Idea* and is currently at work on his second book, *In Search of Common Sense.*

CHRISTOPHER JOYCE is the U.S. editor of *New Scientist* magazine. He recently completed a book about forensic anthropology called *Witnesses from the Grave: The Stories Bones Tell* with coauthor Eric Stover.

MICHIO KAKU is a professor of physics at City University of New York. He is coauthor (with Jennifer Trainer) of *Beyond Einstein* and author of *Introduction to the Superstring,* as well as the forthcoming *Hyperspace.*

HAROLD KLAWANS is professor of neurology and pharmacology at Rush University in Chicago. He is the author of *Toscanini's Fumble* and *The Medicine of History.*

JOHN KOTRE is a professor of psychology at the University of Michigan, Dearborn, and coauthor (with Elizabeth Hall) of *Seasons of Life.* He has published four books on life-historical subjects, including *Outliving the Self: Generativity and the Interpretation of Lives.*

LAWRENCE M. KRAUSS is an associate professor of physics and astronomy at Yale University and the author of *The Fifth Essence: The Search for Dark Matter in the Universe.* He is currently completing a second book entitled *Fear of Physics.*

THOMAS LEVENSON is a producer of "Nova" for WGBH in Boston. He is the author of *Ice Time* and of a forthcoming history of scientific and musical instruments.

ALAN LIGHTMAN is a professor of science and writing and senior lecturer in physics at MIT, and he is also on the staff of the Harvard–Smithsonian Center for Astrophysics. He is coauthor of *Origins: The Lives and Worlds of Modern Cosmologists,* which won the Association of American Publishers award for the best book on physical science in 1990.

ROBERT MARCH, a professor of physics at the University of Wisconsin, Madison, is author of *Physics for Poets* and numerous articles. He is a two-time winner of the Science Writing Award from the American Institute of Physics.

LAURENCE A. MARSCHALL is a professor of physics at Gettysburg College in Pennsylvania and the author of *The Supernova Story*. He is a contributing editor for *Smithsonian Air and Space* and writes a regular column, "Books in Brief," for *The Sciences*.

HAROLD MOROWITZ is a Clarence Robinson professor of biology and natural philosophy at George Mason University. His latest books are *The Thermodynamics of Pizza* and *Beginnings of Cellular Life*.

DENNIS OVERBYE is the author of *Lonely Hearts of the Cosmos: The Scientific Quest for the Secret of the Universe,* which was nominated for the National Book Critics Circle award for nonfiction and won the American Institute of Physics Science Writing award. He is a contributing essayist for *Time* magazine.

PETER RADETSKY is a contributing editor for *Discover* magazine and teaches in the Science Communication Program at the University of California at Santa Cruz. He is the author of *The Invisible Invaders: The Story of the Emerging Age of Viruses,* which was named to the American Library Association's 1992 Best Books List.

WILLIAM H. SHORE is the founder and executive director of Share Our Strength, a nonprofit hunger relief foundation in Washington, D.C.

JUDITH STONE, a contributing editor for *Discover* and *Glamour,* is the author of *Light Elements: Essays on Science from Gravity to Levity*.

JAMES TREFIL is a Clarence Robinson professor of physics at George Mason University. His most recent book is *1001 Things You Should Know About Science*. He is also coauthor, with Robert M. Hazen, of *Science Matters: Achieving Scientific Literacy*.

HANS CHRISTIAN VON BAEYER is a professor of physics at the College of William and Mary, where he has been since 1968. In 1979 he won the science writing award of the American Institute of Physics, and in 1991 his essays for the *Sciences* won a National Magazine Award. He is the author of *Taming the Atom: The Emergence of the Visible Microworld.*

ROBERT WRIGHT is a senior editor of the *New Republic* and the author of *Three Scientists and Their Gods: Looking for Meaning in an Age of Information.* He was formerly a senior editor of *The Sciences,* where he wrote the column "The Information Age." He was winner of the 1986 National Magazine Award for Essay and Criticism.

A. ZEE holds a joint appointment as professor of physics at the University of California and at the National Institute for Theoretical Physics in Santa Barbara. He is the author of *Fearful Symmetry, Swallowing Clouds,* and *An Old Man's Toy,* which was nominated for a Pulitzer Prize.

PERMISSIONS